SpringerBriefs in Architectural Design and Technology

Series Editor

Thomas Schröpfer, Architecture and Sustainable Design, Singapore University of Technology and Design, Singapore, Singapore

Indexed by SCOPUS

Understanding the complex relationship between design and technology is increasingly critical to the field of Architecture. The *Springer Briefs in Architectural Design and Technology* series provides accessible and comprehensive guides for all aspects of current architectural design relating to advances in technology including material science, material technology, structure and form, environmental strategies, building performance and energy, computer simulation and modeling, digital fabrication, and advanced building processes. The series features leading international experts from academia and practice who provide in-depth knowledge on all aspects of integrating architectural design with technical and environmental building solutions towards the challenges of a better world. Provocative and inspirational, each volume in the Series aims to stimulate theoretical and creative advances and question the outcome of technical innovations as well as the far-reaching social, cultural, and environmental challenges that present themselves to architectural design today. Each brief asks why things are as they are, traces the latest trends and provides penetrating, insightful and in-depth views of current topics of architectural design. *Springer Briefs in Architectural Design and Technology* provides must-have, cutting-edge content that becomes an essential reference for academics, practitioners, and students of Architecture worldwide.

Gabriele Bernardini · Elena Cantatore ·
Fabio Fatiguso · Enrico Quagliarini

Terrorist Risk in Urban Outdoor Built Environment

Measuring and Mitigating via Behavioural Design Approach

Gabriele Bernardini
Department of Construction, Civil
Engineering and Architecture
Università Politecnica delle Marche
Ancona, Italy

Elena Cantatore
Department of Civil, Environmental, Land,
Building Engineering and Chemistry
Politecnico di Bari
Bari, Italy

Fabio Fatiguso
Department of Civil, Environmental, Land,
Building Engineering and Chemistry
Politecnico di Bari
Bari, Italy

Enrico Quagliarini
Department of Construction, Civil
Engineering and Architecture
Università Politecnica delle Marche
Ancona, Italy

ISSN 2199-580X   ISSN 2199-5818 (electronic)
SpringerBriefs in Architectural Design and Technology
ISBN 978-981-97-6964-3   ISBN 978-981-97-6965-0 (eBook)
https://doi.org/10.1007/978-981-97-6965-0

This Springer imprint is published by the registered company Springer Nature Singapore Pte Ltd.
The registered company address is: 152 Beach Road, #21-01/04 Gateway East, Singapore 189721, Singapore

If disposing of this product, please recycle the paper.

To Fabio and Martina, my safe harbor
To my wife Federica, my life, and to my
daughter Amelia, my future

Acknowledgments

This research was funded by the Italian Ministry of Education, University, and Research (MIUR) Project BE S^2ECURe—(make) Built Environment Safer in Slow and Emergency Conditions through behavioural assessed/designed Resilient solutions (grant number: 2017LR75XK). The project website is https://www.bes2ec ure.net.

Authors' Contribution

This book is written by Bernardini Gabriele, Cantatore Elena, Fatiguso Fabio and Quagliarini Enrico. Bernardini Gabriele (corresponding Author) and Cantatore Elena dealt with conceptualization, writing (original draft preparation), data curation, formal analysis, investigation and visualization. Fatiguso Fabio and Quagliarini Enrico dealt with review, editing and supervision, In particular, the contribution of each author is reported for every chapter of the book in the following.

Chapter 1: Introduction- Cantatore Elena, Bernardini Gabriele, Quagliarini Enrico, Fatiguso Fabio.

Chapter 2: Terrorist Risk in Urban Outdoor Built Environment: Influencing Factors and Mitigation Strategies - Cantatore Elena, Fatiguso Fabio.

Chapter 3: User Behaviour in Terrorist Acts to Model the Evacuation in Outdoor Open Areas- Bernardini Gabriele, Quagliarini Enrico.

Chapter 4: Measuring and Improving the Resilience of Outdoor Open Areas Against Terrorist Acts: A Behavioural Design Approach - Bernardini Gabriele, Cantatore Elena.

Chapter 5: A Case Study Application: Vittorio Veneto Square in Matera, Italy - Cantatore Elena, Bernardini Gabriele.

Chapter 6: Conclusions and Perspectives- Bernardini Gabriele, Cantatore Elena, Fatiguso Fabio, Quagliarini Enrico.

Contents

1 Introduction 1
  1.1 The Principles of the Terroristic Phenomenon
     for Understanding the Threat in the Outdoor Open Areas 1
  1.2 The Security of Cities, the Human Factor, and the Terrorism
     Threat 5
  References 7

2 Terrorist Risk in Urban Outdoor Built Environment:
  Influencing Factors and Mitigation Strategies 11
  2.1 Terrorist Threat in the European Urban Built Environment:
     Understanding Levels of Riskiness in Outdoor Open Areas
     Using Risk Matrix 11
    2.1.1 The Discretization of the Terroristic Phenomenon
       in the Outdoor Open Areas Within the GTD Database 12
    2.1.2 The Frequentistic Analysis of the Terroristic
       Phenomenon in Western Europe: From the Built
       Environment to the Outdoor Open Area Scale 14
  2.2 Secure Urban Built Environment Prone to the Terrorism
     Threat: The Risk Mitigation and Reduction Strategies 18
    2.2.1 Classification of Risk Mitigation and Reduction
       Strategies in the Built Environment: An International
       Overview 19
    2.2.2 The Sustainability of Risk Mitigation and Reduction
       Strategies in the Built Environment 21
  2.3 Factors Affecting the Terroristic Risk in the Outdoor Open
     Areas for the Most Recurrent Attack Typologies 26
  References 31

**3   User Behaviour in Terrorist Acts to Model the Evacuation
     in Outdoor Open Areas**   35
     3.1   Understanding and Simulating User Behaviours in Terrorist
           Acts to Support Risk Assessment and Mitigation   35
     3.2   User Behaviour in Terrorist Acts   36
     3.3   Summary of Main Motion Quantities in Terrorist Evacuation   37
     3.4   Towards an Evacuation Model for Terrorist Acts Simulation
           in the Urban Outdoor Open Areas   44
           3.4.1   Main Modelling Issues of the OA   47
           3.4.2   Main Modelling Issues of the Attackers   49
           3.4.3   Main Modelling Issues of the Users   51
     References   55

**4   Measuring and Improving the Resilience of Outdoor Open
     Areas Against Terrorist Acts: A Behavioural Design Approach**   59
     4.1   From Risk Scenarios to Risk Assessment and Mitigation
           in Outdoor Open Areas   59
     4.2   Measure the Risk Assessment of Outdoor Open Areas
           to Provide Possible Attack Points in Real Case Study   60
     4.3   Methods for Time-Dependent Assessment of Users-Related
           Factors   67
     4.4   Simulation-Based Indicators   75
     4.5   Mitigation and Preventive Strategies Towards Effectiveness
           and Outdoor Open Areas Compatibility   79
     References   88

**5   A Case Study Application: Vittorio Veneto Square in Matera,
     Italy**   93
     5.1   The Case Study: Vittorio Veneto Square in Matera, Italy   93
     5.2   Risk Assessment of OAs to Provide Possible Attack Points:
           Pre-Retrofit Scenarios   96
     5.3   Mitigation Strategies Identification: Post-Retrofit Scenarios 101
     5.4   Time-Dependent Assessment of User-Related Factors 105
     5.5   Simulation Scenarios and Results 111
     References 115

6   Conclusions and Perspectives 117
     6.1   Outdoor Open Areas and Terrorist Acts: How Behavioural
           Design Could Support Risk Assessment and Mitigation? 118
     6.2   Perspectives in Research and Practice 120
           6.2.1   Risk Matrix 121
           6.2.2   Behavioural Modelling and Simulation 122
           6.2.3   Scenario Creation and KPI-Based Risk Assessment 122
           6.2.4   Risk Mitigation 124
     References 125

Notations

Symbol/ acronym	Short definition [main chapters of relevance]	Unit of measure
BE	Built Environment [all]	
CBE	Class of Built Environmen [2, 4, 5]	
CR	Casualty ratio [4, 5]	[−]
F	CBE composed of public uncovered un-built areas, squares, and streets [2, 4, 5]	
F_B	OutBE class referring to public buildings with entertainment uses in the context of CBE F [2, 4, 5]	
F_D	OutBE class referring to representative (symbolic) or strategic buildings in the context of CBE F [2, 4, 5]	
FN95	Normalized flows at the 95th percentile of arrived users [4, 5]	[−]
E	Exposure [4, 5]	
GS	Available gross surface of a given intended use i [4, 5]	[m^2]
H	Hazard related to terrorist event [4, 5]	
IE_t	Impact of an event in the OA on the whole population at a given time t [4, 5]	[−]
KPIs	Key performance indicators [all]	
NA	Not-arrived users' ratio [4, 5]	[−]
NR	Non-residents users [4, 5]	
$NU_{t, i}$	Number of users of a given intended use i in the OA [4, 5]	[persons]
NU_t	Total number of users in the OA [4, 5]	[persons]
NUn_t	Users' normalized number at a given time t [4, 5]	[persons]
OL_i	Quick occupant load of a given intended use i [4, 5]	[persons/ m^2]
OA	Outdoor Open Area [all]	
OO	Only Outdoor users [4, 5]	
PO	Prevalent outdoor users [4, 5]	

(continued)

(continued)

Symbol/ acronym	Short definition [main chapters of relevance]	Unit of measure
PN	Normalized number of physical contacts among the users [4, 5]	[−]
PV	Percentage Variation of a given KPI [4, 5]	[%]
RMRS	Risk Mitigation and Reduction Strategies [1, 2, 4, 5]	
SoR	Space of relevance [4, 5]	
T2	Armed assault [2, 4, 5]	
T3	Bombing attack [2, 4, 5]	
TN95	Normalized evacuation time at the 95th percentile of arrived users	
TP	Terrorism principles [1, 2]	
UOd	Overall users' outdoor density in outdoor at a given time t [4, 5]	[persons/ m^2]
OutBE	Outdoor BE, thus referring to the outside of the buildings having a direct correlation with facing squares and streets [2, 4, 5]	

Chapter 1
Introduction

Abstract Terrorist phenomenon implies complex risks for the urban built environment (BE), due to the combination of perpetrator behaviour, user reaction to possible attacks, and the characterizing features of the BE itself. Among possible scenarios which can attract terrorist acts, outdoor Open Areas (OAs) surely represent critical conditions especially since they are ideal "soft targets". On one side, OAs can be affected by (over)crowding, as well as can have a symbolic value due to the intended uses hosted outdoors and in the facing buildings. On the other side, OAs are also generally characterized by non-structured protection measures due to the possibility to host public, contrarily to "hard targets", such as government buildings or critical infrastructures, where restricted access areas, control systems, and security strategies are widely implemented. This chapter traces the principles for understanding terroristic phenomenon in OAs, and provides basic insights to move from the phenomenology of terrorist acts to the definition of Risk Mitigation and Reduction Strategies according to guidelines and normative framework. The role of user behaviour in such sudden-onset emergencies is also discussed by underlining the connection between the terrorist act, the OAs features and the implemented solutions, since these events can also generally imply the activation of evacuation as one of the most effective protection measures to increase users' safety levels.

Keywords Terrorism · Outdoor open areas · Urban built environment · Risk assessment · Risk mitigation and reduction strategies · Behavioural design

1.1 The Principles of the Terroristic Phenomenon for Understanding the Threat in the Outdoor Open Areas

Terrorism is presently associated with nationalist claims rooted in extremist ideologies arising from political or religious disparities [1, 2]. These characteristics are intricately tied to the human dimension of the threat, posing challenges in parametrizing these events. As the term implies, acts of terrorism are strategically planned to instil

© The Author(s) 2025
G. Bernardini et al., *Terrorist Risk in Urban Outdoor Built Environment*,
SpringerBriefs in Architectural Design and Technology,
https://doi.org/10.1007/978-981-97-6965-0_1

terror, fear, and disorientation. Additionally, terrorist violence exhibits two key char-acteristics: a material function causing immediate physical damage and a symbolic function supporting the concept of terror on a large scale, impacting both the physical dimension of the built environment (BE) and the human dimension of its users. These characteristics are relevant especially when relating to the urban BE, where public spaces (that can be generally associated with outdoor Open Areas—OAs, such as streets, squares, urban parks, and other un-built areas in the urban fabric [3]) can be affected by dynamics in users' attraction over space and time depending on social issues, and which can host different functions with a high level of attack desirability by perpetrators [4–6].

The domains of "threat" and "disaster" concerning terrorism intersect with the management of critical natural events, albeit with five macro differences identified [7–11]:

- Firstly, the significance of damage in terms of casualties and targets in terrorist attacks can have major effects when compared to natural phenomena, espe-cially considering the extension of the areas involved (e.g., the attack in Madrid in 2004 involved different train stations to the bombing attack, causing about 200 victims and about 2000 casualties and the 2009 earthquake in the territory of italian city of L'Aquila, causing about 300 victims and about 1600 casualties).
- Secondly, the choice of target locations in terrorism is driven by human will, aiming at maximizing terror, unlike natural events influenced by statistical and probabilistic factors.
- The intentional nature of perpetrator acts in terrorism contrasts with the predictability of natural events.
- The psychological impact of terrorism can surpass that of natural disasters due to the deliberate nature of violent actions.
- Lastly, disaster mitigation approaches vary significantly between terrorism and natural disasters; nevertheless, while best practice sharing is a fundamental tool for natural hazards, related presentation and dissemination for terroristic events seem to be limited for security reasons, and mainly focused on the discussion of technologies.

While terrorism is not a new phenomenon, its contemporary significance is closely associated with the 9/11 attack in the U.S.A., characterized by symbolism, high casualties, and intricate planning [12]. Recent events in Europe have heightened the attention, emphasizing urban resilience as a strategy to enhance the physical robustness of the BE, aligning with efforts related to natural disasters and supported by increased funding for security projects. Despite major studies and applications on terrorism assessment in cities originating from the USA, European attention to the phenomenon is more recent, prompted by attacks in Madrid (2004) and London (2005), leading to the development of national regulations for regional analysis and management of the threat, particularly in crowded, political, religious, sensitive, and public places. It is the case of the Italian regulations,[1] issued after the tragic events in Torino in 2017 where a false terrorist alarm caused a rapid evacuation of Piazza San Carlo during a public event, and the German experience in managing the security for mass gathering events with organizers.[2] These efforts have been then supported, at the international level, by the definition of guidelines and white books to support risk assessment and mitigation, having a special focus on public space and on architectural and urban design issues (e.g., at the European level, please compare with [4]).

The complexity of terrorism risk assessment revolves around three primary factors: defining the threat, establishing principles, and incorporating multidisciplinary perspectives.

Comprehensive encyclopaedias on terrorism indicate the absence of a universally accepted definition for this phenomenon, emphasizing the localized nature of defining terrorism in national and international regulations [13, 14]. However, three simultaneous key aspects are crucial in characterizing a terrorist act: the perpetration of violent actions aimed at causing fatalities, typically carried out by an individual or an organized group with a coordinated intent for violence, and the selection of symbolic or highly public targets. Consequently, the challenge in defining terrorism represents the initial layer of complexity in risk assessment.

The Global Terrorism Database (GTD)™ stands out as a coherent resource for collecting and managing terrorism-related events.[3] Developed by the National Center for the Study of Terrorism and Responses to Terrorism (START)[4] at the University of Maryland, GTD employs specific terrorism "characters" and "criteria of cruel acts" for effective event identification [15, 16].

Additionally, understanding the logical criteria underlying the terrorism threat is a subject of discussion. In this sense, G. Woo's work [7] represents one of the most significant researches on this topic and highlights the distinctive principles, which have been synthesized into four main terrorism principles (TPs):

TP.1.   The impact factor relates to the concept of maximizing the terrorist attack. This principle can be divided into two macro-categories:

> TP.1.1-Macro-terror is characterized by the reduction of the frequency due to the complexity of attack planning and execution weapons.

---

[1] Ministero dell'Interno, Modelli organizzativi e procedurali per garantire alti livelli di sicurezza in occasione di manifestazioni pubbliche—National Regulation, Roma, 18th July 2018 Available online at http://www.interno.gov.it/it/amministrazione-trasparente/disposizioni-generali/atti-generali/atti-amministrativi-generali/circolari/circolare-18-luglio-2018-modelli-organizzativi-e-procedurali-garantire-alti-livelli-sicurezza-occasione-manifestazioni-pubbliche (last access: 26/02/2024).

[2] Sicherheit öffentlicher Veranstaltungen, Richtlinie zur Erstellung eines Sicherheitskonzeptes, Stadt Münster – national guidelines for the management of mass gathering event - 24th May 2017 available online at https://www.stadt-muenster.de/fileadmin/user_upload/stadt-muenster/32_ordnungsamt/pdf/richtlinien_sicherheitskonzepte2017-05.pdf (last access: 26/02/2024).

[3] https://www.start.umd.edu/gtd/ (last access: 26/02/2024).

[4] https://www.start.umd.edu/ (last access: 26/02/2024).

TP.1.2-Micro-terror is characterized by less management complexity and a high probability of repeatability.

TP.2. The "Publicity Impact is Key to Targeting" highlights the perpetrator's need to maximize media repercussion.

TP.3. Inter-dependence and replacement of targets in compliance with the principle according to which "terrorists will attack the softer of two similarly attractive targets". This principle can be divided into two macro-categories related to protection systems:

TP.3.1-Hard targets, such as government buildings or military headquarters, concern buildings characterized by a system of active or passive protection technologies, regardless of the probability of occurrence. Professionals and relevant political, religious, or media figures belong to this class.

TP.3.2-Soft targets, including subways, pubs, other public spaces as well as vulnerable sites without any type of defence measure against these phenomena. Considering the human relevance aspect, it refers to the community, gathered in extensive urban areas, lacking effective protection systems from attacks.

TP.4. The characterization of terrorist weaponry, relating to the criterion of minimizing resistance, facilitates the evaluation of the level of threat and the equipment type used by the perpetrator. The same prefers traditional and easily available weapons (guns and explosives).

Thus, TPs highlight relevant aspects of the attack goals which include the maximization of the attack impact, the importance of publicity for targeting, the interdependence and replacement of targets, and the characterization of terrorist weaponry.

Despite this analysis of the human phenomenon, the multidimensionality of the terrorism threat is explored in the literature through mono-thematic and detailed studies, covering simulations of human behaviour, economic analyses of losses, countermeasures, and specific attack types about critical urban infrastructures or specific case studies [17–30]. On the other hand, there is a gap in the general risk assessment and multi-temporal management of urban areas, leaving certain aspects unexplored.

Among public spaces, OAs surely represent a paramount class within the "soft targets" (compared to TP.3.2) in the urban BE [4]. In fact, in OAs, micro-terror may fully describe the goal of perpetrators in violent actions, while their higher proneness in suffering the re-iteration of actions can be related to the lower levels of protection that usually characterize such places in daily use.

On the other hand, the concept of OAs as a complex system of buildings, users, and infrastructures, serving and interacting within the perceived urban un-built area, requires to be analysed in depth, trying to understand which and how uses and services contribute to enhance or reduce the proneness of places, while physical features are addressed to understand the inherent vulnerabilities and use to evaluate the potential exposure of the violent events.

Starting from this, the following Chap. 2 presents the results of the phenomenological analysis of the terrorist events in European cities, trying to understand the relevance of the OAs, both as un-built areas and as a system, as soft targets within the urban extension, considering the GTD database.

1.2 The Security of Cities, the Human Factor, and the Terrorism Threat

Coherently with the main sustainable goals of the "secure and safe cities", existing literature and experiences applied within urban BE to mitigate and reduce the hazards and effects of terroristic events offer the opportunity to understand the interrelations between the physical environment and the human factor during the events. To this end, Risk Mitigation and Reduction Strategies (RMRSs) have been already observed and translated into regulations and guidelines which serve to guide urban policymakers in guaranteeing urban security and users' safety.

Considering major international guidelines, RMRSs can operate in two distinct modes and timeframes [31–34]:

- Pre-event, aiming to prevent, detect, and delay emergency conditions through preventive measures or management procedures implemented by stakeholders and law enforcement agencies (LEAs);
- Throughout the violent act, where strategies must minimize casualties and facilitate evacuation with the support of LEAs and the defensive organization of the BE layout, guiding individuals to adopt safe behaviours during emergency phases.

Indeed, these issues should be correlated not only with their impact on the target desirability by the perpetrators, but also with the users hosted in the BE, who can adopt different behaviours depending on the stressors they are facing, as well as on the level of protection and safety perception given by the BE itself, the effects of the attack, and the implemented RMRSs [6, 32, 35–37]. As suggested by previous works for different kinds of emergencies affecting the BE (e.g. fires [38]) and, in particular, the OAs (e.g., earthquakes [39]), including the "human factor" in risk assessment and mitigation can effectively support the development of RMRs also in respect to terrorist acts [35]. The behavioural design approach moves in this direction, considering the analysis of users' exposure, vulnerability, and behaviours in emergency conditions as the key element to support such tasks.

In particular, the response to the violent act implies the interactions between the users and the perpetrators considering both the attack itself (e.g., users killed or wounded by the terrorists) and the evacuation process (i.e., users move far from the attack area to restore safety and protect themselves from the terrorists). User behaviour hence includes risk perception before the event (which can affect permanence tasks in the BE), and motion tasks during the emergency (i.e., motion speed, path selection, run-hide-cover, and also fight behaviour against the perpetrators), and previous studies suggested how they can be considered as consistently different from those noticed in other kinds of emergencies (e.g., fires, earthquakes) or general purpose emergencies and evacuation [35] (compare with Chap. 3). Understanding and modelling the user behaviour in terrorist acts can support the definition of such RMRSs, thus defining the basis for implementing a complete behavioural design approach to terrorist acts. User-oriented assessment can be then combined with issues related to the specificities of the perpetrators' "modus operandi", the liveability of the OAs, the applicability to specific contexts in terms of morphology, identity features (e.g., historical OAs), and intended uses (e.g., public spaces also used for mass gatherings) [36].

In that sense, it is worth noting that the overall picture of RMRSs is already well-defined for "hard targets" and specifically for special buildings or places, such as government buildings, critical infrastructure, and police stations. This cannot exclude them from their assessment for the application in "soft target" contexts, including OAs but can help the understanding of the emergency phase where evacuation safety regulations can adhere to common standards applicable to both hard and soft targets [31, 40, 41]. The study of such guidelines may support the comprehension of good practices in the emergency phase, and also the understanding of how and which physical elements and properties should be included in risk assessment and risk mitigation design for "soft targets", aiming at a sustainable and effective design of strategies and solutions [37, 41–45].

On the other hand, the design of solutions should consider sustainable applicability in real places, taking advantage of redundancy, adaptability, coordination, and costs as determinants to compare RMRSs and to evaluate how RMRSs combined applications can be implemented [32, 46–49].

In that sense, combining the international experiences about risk mitigation and the phenomenological analysis of the European terrorism threat, fast methods to determine the class of risk for real OAs can represent key tools to support local administrations and their designers/technicians, assuming that they can also have a low level of knowledge on the matter (compare with methods defined in Chap. 4, mainly declined for Italian case studies). Specifically, all the properties and elements that may interact in the risk assessment should be jointly considered in order to provide a tool to compare the riskiness of real OAs and to determine the possible attack points within a place.

Moreover, the characterization of RMRSs in systems of effective, compatible, and redundant strategies can support the choice of a well-designed solution for real OAs, considering the main results of the behavioural-based assessment during the events in pre- and post-designed scenarios. To this end, according to the behavioural design approach [35], simulation-based methods can provide useful insights into the specific dynamics affecting the event with respect to the user response to the perpetrators' actions [37, 50]. Nevertheless, in view of the similarities and differences in user behaviour during terrorist acts and other emergencies [35, 41], specific modelling tools should be developed according to experimental data (see Chap. 3), and then risk indicators should be defined to evaluate the impacts of certain attack affects and user behaviour on safety levels (see Chap. 4). Finally, applications to real-world case studies (see Chap. 5) can provide insights into the reliability and capability of this behavioural design approach.

References

1. Bahgat K, Medina RM (2013) An overview of geographical perspectives and approaches in terrorism research. Perspect Terrorism 7:38–72
2. Marineau J, Pascoe H, Braithwaite A et al (2020) The local geography of transnational terrorism. Confl Manag Peace Sci 37:350–381
3. Jayakody RRJC, Amarathunga D, Haigh R (2018) Integration of disaster management strategies with planning and designing public open spaces. Procedia Eng 212:954–961. https://doi.org/10.1016/j.proeng.2018.01.123
4. The European Commission (2022) Security by design: protection of public spaces from terrorist attacks
5. Quagliarini E, Bernardini G, Romano G, D'Orazio M (2023) Users' vulnerability and exposure in public open spaces (squares): a novel way for accounting them in multi-risk scenarios. Cities 133:104160. https://doi.org/10.1016/j.cities.2022.104160
6. Kalvach Z et al (2016) Basics of soft targets protection—guidelines (2nd version). Prague
7. Woo G (2015) Understanding the principles of terrorism risk modeling from charlie hebdo attack in Paris. Defence Against Terrorism Review-DATR 7:1–11
8. Li Piani T (2018) Progettazione strutturale e funzione sociale dello spazio (quale) vulnerabilità e soluzione al terrorismo urbano. Sicurezza, terrorismo e società. INTERNATIONAL JOURNAL—Italian Team for Security, erroristic Issues and Managing Emergencies (in italian; ISSN: 2421–4442) 7–15
9. Godschalk DR (2003) Urban hazard mitigation: creating resilient cities. Nat Hazard Rev 4:136–143
10. Turégano-Fuentes F, Pérez-Díaz D, Sanz-Sánchez M, Alonso JO (2008) Overall asessment of the response to terrorist bombings in trains, Madrid, 11 March 2004. Eur J Trauma Emerg Surg 34:433
11. Alexander D, Magni M (2013) Mortality in the l'aquila (central Italy) earthquake of 6 April 2009. PLoS currents 5. https://doi.org/10.1371/50585b8e6efd1
12. Averill JDD, Mileti DSS, Peacock RDD, et al (2005) Federal building and fire safety investigation of the world trade center disaster: occupant behavior, egress, and emergency communications (NIST NCSTAR 1–7). U.S. Government printing office, Washington, D.C
13. Combs CC, Slann MW (2009) Encyclopedia of terrorism. Infobase Publishing
14. Crenshaw M, Pimlott J (2019) Encyclopedia of world terrorism. Routledge
15. National Consortium for the Study of Terrorism and Responses to Terrorism (START) Global Terrorism Database (GTD) l. www.start.umd.edu/gtd/search/. Accessed 1 Dec 2019
16. National Consortium for the Study of Terrorism and Responses to Terrorism (START) (2019) Global Terrorism Database Codebook: Inclusion Criteria and Variables
17. Leweling TA, Nissen ME (2007) Defining and exploring the terrorism field: toward an intertheoretic, agent-based approach. Technol Forecast Soc Chang 74:165–192
18. Geng X, Li G, Ye Y, et al (2006) Abnormal behavior detection for early warning of terrorist attack. In: Australasian joint conference on artificial intelligence. Springer, pp 1002–1009
19. Davis PK, Perry WL, Brown RA, et al (2013) Using behavioral indicators to detect potential violent acts

20. Doss K, Shepherd C (2015) Active shooter: preparing for and responding to a growing threat. Butterworth-Heinemann
21. Mueller J, Stewart MG (2011) Balancing the risks, benefits, and costs of homeland security. Homeland security affairs 7
22. Thöns S, Stewart MG (2019) On decision optimality of terrorism risk mitigation measures for iconic bridges. Reliab Eng Syst Saf 188:574–583
23. Stewart MG (2017) Risk of progressive collapse of buildings from terrorist attacks: are the benefits of protection worth the cost? J Perform Constr Facil 31:4016093
24. Richardson HW, Pan Q, Park J, Moore JE (2015) Regional economic impacts of terrorist attacks, natural disasters and metropolitan policies. Springer
25. Zaidi MK (2007) Risk assessment in detection and prevention of terrorist attacks in harbors and coastal areas. In: Environmental Security in Harbors and Coastal Areas. Springer, pp 309–316
26. De Cillis F, De Maggio MC, Pragliola C, Setola R (2013) Analysis of criminal and terrorist related episodes in railway infrastructure scenarios. J Homel Secur Emerg Manage 10:447–476
27. Frolov KV, Baecher GB (2006) Protection of civilian infrastructure from acts of terrorism. Springer Science and Business Media
28. Shvetsov AV, Shvetsov MA (2019) A fast-track method for assessing the risk of a terrorist attack on transportation facilities. Euro J Secur Res 4:265–271
29. Lehr P (2018) Counter-terrorism technologies: a critical assessment. Springer
30. Janzon B, Forsén R (2008) Threats from terrorist and criminal activity and risk of dangerous accidents—resistance and vulnerability of the urban environment and ways of mitigation. In: Resilience of Cities to Terrorist and other Threats. Springer, pp 3–36
31. Home Office in partnership with the Department for Communities and Local Government (2012) Crowded places: the planning system and counter-terrorism
32. Coaffee J, O'Hare P, Hawkesworth M (2009) The visibility of (in) security: the aesthetics of planning urban defences against terrorism. Secur Dialogue 40:489–511
33. Karlos V, Larcher M, Solomos G (2018) Review on soft target/public space protection guidance. Publications Office of the European Union, Luxemburg
34. Cuesta A, Abreu O, Balboa A, Alvear D (2019) A new approach to protect soft-targets from terrorist attacks. Saf Sci 120:877–885. https://doi.org/10.1016/j.ssci.2019.08.019
35. Bernardini G, Quagliarini E (2021) Terrorist acts and pedestrians' behaviours: first insights on European contexts for evacuation modelling. Saf Sci 143:105405. https://doi.org/10.1016/j.ssci.2021.105405
36. Quagliarini E, Fatiguso F, Lucesoli M et al (2021) Risk reduction strategies against terrorist acts in urban built environments: towards sustainable and human-centred challenges. Sustainability 13:901. https://doi.org/10.3390/su13020901
37. Li S, Zhuang J, Shen S (2017) A three-stage evacuation decision-making and behavior model for the onset of an attack. Transp Res Part C: Emerg Technol 79:119–135. https://doi.org/10.1016/J.TRC.2017.03.008
38. Kobes M, Helsloot I, de Vries B, Post JG (2010) Building safety and human behaviour in fire: a literature review. Fire Saf J 45:1–11. https://doi.org/10.1016/j.firesaf.2009.08.005
39. Bernardini G, D'Orazio M, Quagliarini E (2016) Towards a "behavioural design" approach for seismic risk reduction strategies of buildings and their environment. Saf Sci 86:273–294. https://doi.org/10.1016/j.ssci.2016.03.010
40. Joint Counterterrorism Assessment Team (JCAT) (2018) Planning and Preparedness Can Promote an Effective Response to a Terrorist Attack at Open-Access Events
41. Zhu R, Lin J, Becerik-Gerber B, Li N (2020) Human-building-emergency interactions and their impact on emergency response performance: a review of the state of the art. Saf Sci 127:104691. https://doi.org/10.1016/j.ssci.2020.104691
42. Bernardini G, Lovreglio R, Quagliarini E (2019) Proposing behavior-oriented strategies for earthquake emergency evacuation: a behavioral data analysis from New Zealand, Italy and Japan. Saf Sci 116:295–309. https://doi.org/10.1016/j.ssci.2019.03.023
43. Gin JL, Stein JA, Heslin KC, Dobalian A (2014) Responding to risk: awareness and action after the September 11, 2001 terrorist attacks. Saf Sci 65:86–92. https://doi.org/10.1016/j.ssci.2014.01.001

44. Alnabulsi H, Drury J, Templeton A (2018) Predicting collective behaviour at the Hajj: place, space and the process of cooperation. Philos Trans Royal Soc B: Biol Sci 373:20170240. https://doi.org/10.1098/rstb.2017.0240

45. Templeton A, Neville F (2020) Modeling collective behaviour: insights and applications from crowd psychology. In: Gibelli L (ed) Crowd dynamics. Springer Nature Switzerland AG, pp 55–81

46. Kılıçlar A, Uşaklı A, Tayfun A (2018) Terrorism prevention in tourism destinations: security forces vs. civil authority perspectives. J Destin Mark Manag 8:232–246. https://doi.org/10.1016/j.jdmm.2017.04.006

47. Federal Emergency Management Agency (2007) FEMA 430: site and urban design for security: guidance against potential terrorist attacks

48. Llewelyn-Davies (Firm) HMP (2004) Safer places: the planning system and crime prevention. Thomas Telford Ltd

49. GCDN Commissioned Research (2018) Beyond Concrete Barriers Innovation in Urban Furniture and Security in Public Space

50. Song Y, Liu B, Li L, Liu J (2022) Modelling and simulation of crowd evacuation in terrorist attacks. Kybernetes. https://doi.org/10.1108/K-02-2022-0260

Chapter 2
Terrorist Risk in Urban Outdoor Built Environment: Influencing Factors and Mitigation Strategies

Abstract Starting from the established and common principles of the terrorism threat in the cities, this chapter presents the results of the phenomenological analysis in Europe and reorganizes the main literature and international experiences in the prevention, mitigation, and management of the threat in the built environment in order to delineate the factors that influence the risk of outdoor Open Areas (OAs) as "soft targets". In fact, if several previous experiences have already investigated the effects of events on people involved, the strategies used by perpetrators and tested mitigative strategies in detailed case studies following an "a posteriori" approach, a unique approach to describe and discuss the risk of OAs seems to be still unexplored. The aim is thus reached by merging two levels of details. I) The assessment of events during the last 20 years in Western Europe allows to understand how (the attack type) and why (which uses affect the likelihood of events) OAs are emergent "soft targets". On the other hand, II) the critical categorization of Risk Mitigation and Reduction Strategies already experimented and regulated in the international panorama helps in highlighting how such soft targets can be physically improved towards resilient parts of the cities.

Keywords Phenomenological analysis · Terrorism risk assessment · Risk mitigation and reduction strategies · Outdoor open areas · Europe

2.1 Terrorist Threat in the European Urban Built Environment: Understanding Levels of Riskiness in Outdoor Open Areas Using Risk Matrix

In the extensive range of risks to which the urban built environment (BE) is exposed, terrorism is classified as a Sudden Onset Disaster (SUOD), caused by human will. Unlike those generated by natural processes, terrorism is driven by the ideology of a political or religious movement and aims to instil fear and destabilize a community through hostile and violent actions carried out on symbolic targets with high media impact. However, while radicalism constitutes the driving force behind the goals of

© The Author(s) 2025

G. Bernardini et al., *Terrorist Risk in Urban Outdoor Built Environment*, SpringerBriefs in Architectural Design and Technology, https://doi.org/10.1007/978-981-97-6965-0_2

violent actions, recent studies on terrorism have highlighted the high variability in the modus operandi of terrorist acts, requiring specific analyses at a macro-territorial scale for understanding events [1]. This variability is not only influenced by pre-existing political, social, and/or cultural relations between the attacking nations and the radicalist matrix but also considers the possibilities of weapons procurement and the feasibility of the attack in relation to specific protection and security measures in place. It is not coincidental that counterterrorism actions operate at the national level with evident variability even on an international scale [2].

On the other hand, as suggested by the latest events in Europe, outdoor Open Areas (OAs) are commonly described as "soft targets" as a consequence of the lower levels of protection usually present in such places [3]. However, when the focus is on the OAs as a system of infrastructures, buildings, un-built areas, and users, it is necessary to understand how their uses can influence the global riskiness of the OAs themselves and as a whole.

Due to these points of discussion and coherently with the common strategies for understanding complex issues, an analysis of previous historical events is required to describe the threat quantitatively. In detail, to support this aim a phenomenological analysis of the violent events is carried out aiming at solv-ing the following goals:

- Determine the most frequent and riskiness scenarios in the urban BE considering homogeneous classes of uses of places and buildings and the weapon types to provide if some uses can alter the inherent level of proneness of OAs (Goal 1—G1)
- Identify the most relevant (in frequency and efficaciousness) weapon types in increasing the global riskiness of places considering the uses of buildings facing the OAs, in order to delineate dominant traces to focus on for OAs (Goal 2—G2)

The analysis of the phenomenon is focused on Western European countries, in view of the significant relevance of the phenomenon and to the recent development of guidelines on the matter [3]. Specifically, the analysis starts from a discretization process of the urban BE and the weapon types applied to recent violent acts in the European territory.

2.1.1 The Discretization of the Terroristic Phenomenon in the Outdoor Open Areas Within the GTD Database

Coherently with the goals of the analysis, the terroristic phenomenon in Western Europe is pursued by means of the analysis of the recorded traumatic occurrences catalogued in the Global Terrorism Database (GTD)™. It is the most complete and extended database for terroristic events worldwide thanks to the interpolation of research actions, taxonomy, and cataloguing activities of the National Center for the Study of Terrorism and Responses to Terrorism (START), established in the University of Maryland. The START activities aim at merging previous databases and enriching them coupling external data (articles, legal documents, etc.) in order to obtain a coherent and structured collection of details. The first attacks that appear in the GTD refer to the period between 1970 and 1997 and were collected by a private security agency, the Pinkerton Global Intelligence Service (PGIS). The digitization process of information, collected by START, continued with the collaboration of the Center for Terrorism and Intelligence Studies (CETIS). It expanded the quantity of information of each attack, beyond 1997. From 2008 to 2011, the data and information search were carried out by the Institute for the Study of Violent Groups (ISVG). The University of Maryland continued the research until 2020, structuring the database coherently with a rigid set of criteria, which involves geographical data (e.g., coordinates, country, city), the date of the terrorist event and details about the cruel events (attack type, type of weapon and numbers, target type, information on perpetrators, causalities). In that sense, the parametrization of criteria and data are the basic conditions to explore the phenomenological analysis of the terroristic events in Europe—and specifically in its western part—adapting them coherently with the goals of the analysis. In fact, the focal point involves the systematic correlation of data pertaining to event frequency and their ensuing consequences, deriving information and details about OAs, in terms of matrices of risk.

However, in order to take adherent information about terroristic events in Western Europe related to the OAs, the target information within the GTD database has been re-elaborated highlighting the events users and space targeted. Specifically, six macro-classes of uses of the built environment (CBE) are derived from the classification within the events recorded in the database, while all the attacks oriented to people are excluded. The process of classification of the BE results in the detailed classes summarized in Table 2.1, where similar uses of buildings can be recognized. In addition, the criterion of the attack type in the database has been preserved in order to discuss the weapon types. In this case, eight typologies are identified and summarized in Table 2.2.

On the other hand, the consequences related to the effects of the violent acts are obtained considering the number of injured persons and victims.

Table 2.1 Details of the classes of built environment (CBE) considered in the parametrization

Code of CBE	Class of built environment
A	Transportation infrastructure (airport, docks, metro, and rail stations)
B	Public buildings with entertainment uses (theatres, museums, bars, restaurants, hotels, shopping centres)
C	Hospitals, schools, universities
D	Representative (symbolic) or strategic buildings
E	Residential buildings and industries
F	Public un-built areas, squares, and streets

Table 2.2 Codification of the attack types coherently with the "attack information" of the GTD

Code	Description
T1	Assassination
T2	Armed assault
T3	Bombing/Explosion
T4	Hijacking
T5	Barricade incident
T6	Kidnapping
T7	Facility/Infrastructure attack
T8	Unarmed assault

2.1.2   The Frequentistic Analysis of the Terroristic Phenomenon in Western Europe: From the Built Environment to the Outdoor Open Area Scale

The applied method for the comprehension of the terrorism phenomenon in Western Europe is underpinned by the construction of risk matrices derived from the combination of:

- The frequentistic probability (P_F), representing the outcomes of event observation (the ratio between the number of occurrences of a specific event type and the total number of events).
- The consequences (C) in terms of damage, evaluated as the cumulative sum of injured individuals and victims, excluding the details about building damages.[1]

Then, P_F and C values are translated into homogeneous classes of "likelihood" and "consequences" considering the medium value of data as central descriptors of the phenomenon and distributing minimum and maximum values within the five classes. Finally, the resulting matrices summarize the levels of likelihood and consequences within defined ranges of P_F and C, offering a rapid reading of common recurrences in the phenomenon. It is in line with an "a posteriori analysis" of a general phenomenon usually used for risk assessment [4, 5].

The phenomenon is thus focused on a representative sample of events. The period of attention is referred to 2001–2020, in order to consider the Twin Tower attack (11th September 2001) as a breaking event towards the current concept of terroristic threat in a significant extension of the period (20 years). The selection of events for the phenomenological analysis is summarized in Fig. 2.1, where the number of events is details, too. Moreover, due to the goals of the phenomenon reading (G1 and G2), the analysis has been specified following two levels of detail:

- The first focuses on the whole set of events that occurred in Western Europe BEs during the selected period ($GTD_{BE2001-2020}$).

---

[1] The GTD database is victim-centred, recording data about people involved and injuries, while neglecting details about physical damages of properties.

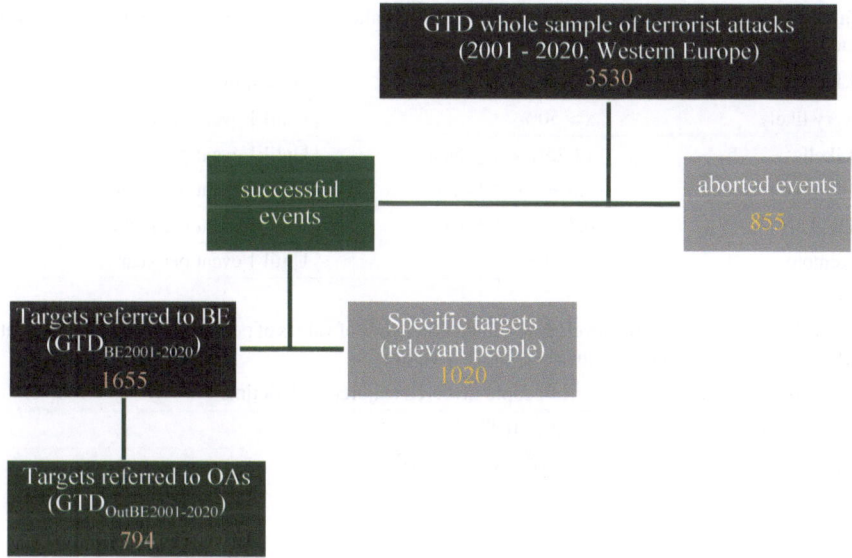

Fig. 2.1 Process of selection of events for the samples, detailed for GTD$_{BE2001-2020}$ and GTD$_{OutBE2001-2020}$

- The second concerns the events that occurred outside the buildings, to relate the inherent proneness of squares and streets to the uses of buildings (GTD$_{OutBE2001-2020}$), which counts approximately 50% of the whole one.

The results of the phenomenon analysis of the BE controlled by attack types and CBEs have shown two main data:

- An attack type is mainly pursued every week in the BE of Western Europe.
- The main value of the consequences for the whole set of violent events counts three people (victims and injured).

Thus, by combining the main values for the identification of the levels of likelihood and consequence (Tables 2.3 and 2.4), the risk matrix has been set up to discuss the terroristic phenomenon in the BE of Western Europe, assuming values from 1 to 5 to each level of likelihood and consequence and calculating the final classes of risks (P_F x C) in coherence with Fig. 2.2.

The resulting risk matrix (Fig. 2.3) highlights four conditions of particular interest. Regarding public open spaces, and thus OAs (CBE F), a medium to high-risk exposure for T2 (armed assault) and T3 (bombing/explosion) attacks is evident, driven by the elevated probability of occurrence (T3) and the generated impact (T2 and T3). Compared to the CBE B, events in this context exhibit high hazard due to prevalent recurrence and effect in combination with T2 and T3 attack types. This is attributed to the intrinsic nature of all public buildings (pubs, museums, etc.) falling under this environmental class, typically characterized by low control levels.

Table 2.3 Likelihood levels determined for classes of values of P_F referred to the period (7300 days) and extended description

Likelihood levels	P_F range	Description
Very likely	x > 50%	Until 1 event per day
Likely	14.25% < x ≤ 50%	Until 1 event per 2 days
Possible	3.3% < x ≤ 14.25%	Until 1 event per week
Unlikely	0.3% < x ≤ 3.3%	Until 1 event per month
Remote	x ≤ 0.3%	Until 1 event per year

Table 2.4 Consequence (C) levels determined for classes of values of people involved in the violent acts, considering the sum of injured people and victims

Consequence levels	n. of people involved (injured and victims)
Extreme (Ex)	$x > 3 \times 10^3$
Major (Ma)	$3 \times 10^2 < x < 3 \times 10^3$
Medium (Me)	$3 \times 10^1 < x < 3 \times 10^2$
Moderate (Mo)	$0 < x \leq 3 \times 10^1$
Minor (Mi)	$x = 0$

Risk Levels	Range of Risk values (P_F x C)
Very high	15 < R1 < 25
High	8 < R2 < 14
Medium	4 < R3 < 7
Low	1 < R4 < 3

Fig. 2.2 Classes of risk levels identified for the BE in Western Europe

		Minor		Moderate		Medium			Major		Extreme		
		Low				Medium							
Remote		All the others		F/T1	F/T4	D/T1 A/T2 C/T2 B/T1 D/T2 D/T3			B/T1 D/T2 D/T3		F/T2	B/T5	1
						C/T3 D/T5 B/T8 D/T7 D/T8 F/T8							
Unlikely		A/T7	F/T7			E/T2 E/T3 E/T7			B/T2	F/T3	A/T3		2
Possible		B/T7		B/T3	D/T3	B/T7					B/T3		3
Likely													4
Very likely													5
		Medium		High		Very High							
		1		2		3			4		5		

Fig. 2.3 Matrix risk of the terroristic phenomenon in BE of Western Europe (reference sample $GTD_{BE2001-2020}$). Cells combine the CBE and attack type classification according to Tables 2.1 and 2.2 (void cells imply that no specific combination is present), respectively, while colours refer to Fig. 2.2

Events occurring in the CBE D (representative or strategic buildings) constitute an intermediate condition with significance in combination with T3 attacks. However, this highlights the greater feasibility of events occurring outside these buildings.

Lastly, airports and railways constitute the CBE A intrinsically more critical due to the potential crowd density, resulting in a high-risk value due to a combination of the number of involved users and the low frequency of events.

However, this corresponds to the highlighting of two fundamental aspects. The first is related to the high susceptibility of open and public spaces characterized by low strategic relevance to undergo an attack. This is in line with the intrinsic definition of soft targets, where confined and non-confined spaces are not easily predisposed to violent acts due to the absence of control and protection systems, regardless of the number of users.

Therefore, CBE A can be excluded from the evaluation, considering the high crowding levels of users and the presence of protection and control systems, which configure it as a hard target. The second aspect is instead related to the specificities of the identified attack locations for CBE D. The use of vehicles as a tool for perpetrating violent acts highlights that the relevance of such events should not only be assessed as acts of high symbolism towards the building and the users within that confined space but also extends to the external environment. In fact, this confirms the necessity of analysing the OAs as a complex system of buildings, infrastructure, open space, and users.

Based on these considerations, the second level of analysis of the phenomenon was led on the reduced sample ($GTD_{OutBE2001-2020}$), following the same frequentistic details (Tables 2.3, 2.4, Fig. 2.3) to determine the matrix of risk for OAs. However, only events that occurred outside of CBEs D and B are considered, being part of OutBE classes, since they are correlated to events that occurred outside the buildings to relate the inherent proneness of squares and streets to intended uses of buildings.

Due to their connection with OAs (CBE F in Table 2.1), these classes are herein considered in a systemic way with F one (F_B, F_D, F). Therefore, the matrix of risk for OAs, summarized in Fig. 2.4, is based on the significance of these three OutBE classes, combined with two main attack types. Specifically, T2 and T3 result in all combinations being classified as high ($F_B/T2$, $F_D/T2$, $F_B/T3$, $F_D/T3$, F/T3) or very high (F/T2) risk levels. The reason for their heightened risk lies in the higher consequence levels, which are directly influenced by the likelihood of these areas being crowded.

In addition, five main results can be summarized for OAs, merging the quantitative results with the principles of the phenomenon (TP—Chap. 1, Sect. 1.1), as follows:

1. F and F_B are found to be more vulnerable than F_D (strategic and symbolic ones) due to their distinct "protection and security systems" that serve as a lesser deterrent (Inter-dependence and replacement of targets, TP.3).
2. The significance of T7 in strategic and symbolic areas reflects the symbolic importance of F_D, where attacks target the environment itself ("Publicity Impact is Key to Targeting" TP.2).

Likelihood Levels	Consequence Levels								
	Minor		Moderate	Medium		Major		Extreme	
	Low					Medium			
Remote	All the others		F_B/T8	F/T4, F_D/T8, F_D/T5	F_B/T1, F_D/T1	F/T8	F_B/T5		1
Unlikely	F/T7	F/T1			F_B/T7	F_D/T7	F_B/T2, F_B/T3, F/T3 | F_D/T2, F_D/T3	F/T2	2
Possible									3
Likely									4
Very likely									5
	Medium		High	Very high					
	1		2	3		4		5	

Fig. 2.4 Matrix risk of the terroristic phenomenon in OAs of Western Europe (reference sample $GTD_{OutBE2001-2020}$). Cells include attack type classification according to Tables 2.1 and 2.2, respectively, while colours refer to Fig. 2.2

3. T2 and T3, being the most frequent attack types, align with the principle that guides the choice of weaponry by terrorists (weaponry characterization TP.4).
4. T2 and T3 represent the most utilized attack types combining the lower level of resistance in perpetrating the violence (TP.1—micro-terror), and consequently, they generate the most significant impacts.
5. The amplification of impact is particularly pronounced in the F_B OutBE Class. In contrast to F_D, which has a higher level of openness due to the need for visibility, the presence of obstacles in F_B can impede escape and decrease overall resilience (TP.3—soft targets).

2.2 Secure Urban Built Environment Prone to the Terrorism Threat: The Risk Mitigation and Reduction Strategies

As introduced in the previous sections, the focus of terrorist risk lies in the will of those planning and executing the event to commit a violent act towards crowds and significant locations. However, considering the complexity of urban BEs, it is evident that all the events impact the crowd that experiences the event.

If the phenomenological analysis supports the comprehension of the violent acts in the pre-event phase, the reading and systematization of the international experiences and regulation framework support the knowledge about the relationship between urban users and the physical place and its elements for their risk reduction, mitigation, and management. Specifically, the analysis and assessment of the so-called Risk Mitigation and Reduction Strategies (RMRS) allow the comprehension of (i) interferences of the physical space and all its elements with the choice phases (mode and location of the attack), (ii) risk reduction, and (iii) emergency management processes.

2.2.1   Classification of Risk Mitigation and Reduction Strategies in the Built Environment: An International Overview

Starting from the analysis of the main national guidelines and American and European regulations for increased sensitivity to the topic, it is possible to classify RMRS according to five specific criteria:

- The type of target (target-oriented strategies) [6, 7], taking into account the classification already introduced between hard and soft targets. Moreover, this criterion distinguishes RMRS based on the varying level of public area accessibility, restricting the perpetrator's proximity to the sensitive target (human-to-event). A secondary division pertains to potential interactions with users, distinguishing between active actions (generating a bi-univocal relation between overarching governance and urban users in prevention processes such as intelligence, active user education, and security surveillance) and passive actions (application of predefined instructions, e.g. regulatory norms, risk communication, urban space redesign).
- The types of attack (attack-oriented) [8, 9], where the definition, selection, and organization of RMRSs may involve simple or complex control and management systems for public space or sensitive buildings, depending on the possible or anticipated modus operandi, as well as their effectiveness concerning space configuration and predisposition to attack.
- The event timings (time-dependent classification) [10, 11], where, coherently with the times of risk, RMRSs are classified with a focus on prevention (pre-event) or management (post-disaster) capabilities, also in relation to potential attack modalities. The perpetrator's operational approach can significantly impact the choice of RMRSs and predict possible human-to-event and event-to-user interactions (e.g. intervention times for video surveillance activities or first responder actions).
- The morphology and nature of the BE [12, 13], consistently with principles outlined in Chap. 1, Sect. 1.1, a terrorist act may target a specific part of public space, necessitating diverse distribution of RMRSs throughout the entire built sector in which the target is situated (e.g., a car bomb attack in areas characterized by varying vehicular accessibility). This includes design strategies for "zones" or "defence areas", recognizing the boundary (a) externally locating all physical barriers and control systems for entire areas characterized by high vulnerability; (b) intermediate, aiming to limit and protect areas or objects within the physical

boundary of OAs; (c) internal, referring to the envelope of the vulnerable, sensitive, or strategic building, or areas within them (core) when identified as primary targets of the attack.

- Physical or managerial purposes [14, 15] according to which, RMRSs can be geared towards risk reduction or emergency event management, focusing on physical interventions in the urban BE and its sub-parts. Additionally, in relation to the first, attention is directed to coordination and management with strategies related to planning, regulation, as well as user education and risk preparedness to maximize effectiveness.

A second level of categorization relates the elements constituting the BE and RMRSs, focusing on the design by users responsible for urban security. Literature and supporting regulations allow the recognition of four macro-classes of RMRS design (S1, S2, S3, S4), comprising specific physical elements in the BE and its layout. These are appropriately integrated with systems for access control, surveillance, and the management of safety and user protection within it. Specifically, these are referred to:

- S1. The design of the physical elements. Perimeter design [S1.1] and secure envelope [S1.2] [12, 15, 16] are addressed for open spaces requiring heightened levels of security and well-being for users; perimeter design incorporates effective mitigative systems evaluated for impact resistance, geometric efficacy concerning accessibility (for T3 attacks), and compatibility with emergency evacuation flows; when the target is confined to a specific element of the built space, the discussion revolves around the building envelope, particularly concerning explosion dynamics.
- S2. The design of BE layout, combining physical elements [S2.1] and layout geometries [S2.2] [15, 16], aims at identifying and creating secure external (standoff) or internal (sheltering) spaces during emergency events by combining physical elements with risk management tools, such as the design of emergency plans [S2.3] [17, 18].
- S3. Pursuing the access control [S3.1] and surveillance [S3.2] [12, 15, 19, 20]; these are predominantly used in managing large events, often combined with perimeter control systems [S1.1], supported by personnel or advanced technologies (body scanners, optical people counting devices, facial recognition in video surveillance). Effective lighting systems [S3.3] [12] are also recognized as necessary to improve visibility and support during emergency evacuations.
- S4. Ensure safety and security management [15, 21–25] by means of several layers of strategies; the use of security personnel [S4.1] as a preventive strategy to deter attacks, support recognition of aggressors, and provide initial aid during and after an attack. This strategy requires broader support, including emergency plan design [S2.3] and, in this context, should be aimed at developing specific issues related to the planning of first aid interventions [S4.2], and their coordination [S4.3]. Finally, the users' involvement [S4.4] through content sharing on various

devices has shown significant benefits in public security management, especially after recent traumatic events in Belgium, France, and Germany.[2]

2.2.2 The Sustainability of Risk Mitigation and Reduction Strategies in the Built Environment

As it is clear, the multiplicity of guidelines and experiences show the complexity of the design of RMRSs in the BE, which have to ensure preventive actions, facilitate the emergency phases, and guarantee good acceptability by urban users. On the other hand, the design of RMRSs has to face multiple levels of sustainability, including the efficacy towards several attack types and the expected ones, the promotion of redundancy in supporting the risk reduction in all the phases of risk (prevention, mitigation, emergency) [6, 20, 26–29].

With these purposes, the following Table 2.5 summarizes the critical evaluation in promoting the sustainable design of RMRSs in the BE, focusing on the relations between strategies and the BE. Specifically, starting from the analysis of the efficacy of RMRSs (S1–S4, see Sect. 2.2.1) with the classification of attack types (T1 to T8, see Table 2.2), and the possible levels of coordination among S-classes of strategies, Table 2.5 shows the levels of applicability (for indoor and outdoor places, for their possible conditions of use) and the features which influence costs.

Another level of assessment and qualification of RMRSs (for classes and sub-classes) can be related to their interrelations with users in the evacuation processes. The human-centred focus is required to define a set of qualitative screening of potentialities and criticalities of RMRS classes useful in simulation analyses. These should consider, near to the human-to-human interactions (both perpetrator to BE users and among BE users), the interference RMRSs-to-human. As it is clear, two levels of details are discussed for RMRS classes [18, 29–31]:

- The potential interference with behavioural issues;
- Their representability in modelling evacuation processes in simulators.

With that aim, Table 2.6 summarizes the critical behavioural design-based analysis of RMRS classes.

As a final remark, the normative and physical sustainability levels of such systems of RMRS require to be merged with the potential exposure levels that affect the emergency and evacuation process, extending the dimension of the matter towards a holistic approach (compare with Chaps. 3 and 4).

---

[2] These European countries have already experimented "educative" initiatives with urban users by means of smartphone applications to communicate real-time the location of events, as well as guidelines for the suggestion for correct behaviours during the violent event. Two main examples are the Belgian virtual platform info-risques.be (available at: https://centredecrise.be/fr/risques-en-belgique), the German KATWARN mobile application (available at: https://www.katwarn.de/en/system.php) and the French guidelines Gérer la Sureté et la Sécurité Des Événements et Sites Culturels [35] (last website access: 26/02/2024).

Table 2.5 Analysis of levels of sustainability for RMRS classes to consider in their design

RMRS	Redundancy with attack typologies	Coordination with other RMRS classes	Adaptable for existing BE	Main application context (intended use; overcrowding)	Factors influencing costs
Design of the physical elements of the BE [S1]					
Safe perimeter [S1.1]	T3	S2.1, S3.1	Adaptable, if punctual installations are used	For hard targets, because of their complexity level	Adopted technologies, BE perimeter length
Secure envelope [S1.2]	T1, T2, T3	S2.2	Usually, they consider new facades, while interventions on openings are more sustainable	For public buildings featuring high crowding levels	Reinforcement typologies for existing openings technology and extension of facades for new constructions
BE layout [S2]					
Standoff [S2.1]	T3	S1.1, S1.2, S3.1	Massive impact is expected when combined with S1.1. Otherwise, adaptable to the existing layout using management actions	Specific for strategic buildings but extendable to soft targets when hosting a high number of visitors	land use costs in guaranteeing the distances, for new constructions In existing BEs, costs concern the space use management
Sheltering [S2.2]	T2, T3, T8	S2.3, S4.2	Adaptable if limited to shelter areas; not compatible when interventions are applied to facades and structures	Single and strategic buildings with something/ someone to protect	Costs are limited only if intervention is applied to existing shelter areas
Emergency layout [S2.3]	all	S1.1, S2.1, S2.2, S3.1, S4.1, S4.2, S4.3	Adaptable for each situation	Adaptable in each event typology	Width of the emergency area and use of BE
Access control and surveillance in the BE [S3]					

(continued)

Table 2.5 (continued)

RMRS	Redundancy with attack typologies	Coordination with other RMRS classes	Adaptable for existing BE	Main application context (intended use; overcrowding)	Factors influencing costs
Access control [S3.1]	T1 to T6	S1.1, S2.1, S2.3, S4.1, S4.3	Adaptable due to the possibility to limit areas (i.e. square perimeter)	Useful for events with significant crowding conditions	Number of installed control points number employed
Security service [S3.2]	T1 to T6	S1.1, S3.1, S3.3, S4.1, S4.2	Adaptable for all the existing BE (including historical) because not invasive installations	Adaptable to all conditions and uses	Width to monitor Adopted technologies
Illumination [S3.3]	T1, T2, T3, T7	S1.1, S3.1	Adaptable for existing (including historical) BEs with possible restrictions at the technological level (e.g. systems integration/ installation)	Adaptable to all conditions and uses	Number of devices operational and maintenance issues
Safety and security management in the BE [S4]					
Security personnel [S4.1]	T1, T2, T3, T4, T5, T6	S1.1, S2.2, S2.3, S3.1, S4.2, S4.3	Adaptable in each condition	Adaptable to all conditions and uses	Building dimensions and floors In mass gatherings, event area extension and number of participants
First aid [S4.2]	all	S.2.2, S2.3, S4.1, S4.3	Adaptable in each condition	Mandatory for mass gatherings and in hard targets of the BE	Low costs by considering the direct possibility of saving lives

(continued)

Table 2.5 (continued)

RMRS	Redundancy with attack typologies	Coordination with other RMRS classes	Adaptable for existing BE	Main application context (intended use; overcrowding)	Factors influencing costs
Coordination [S4.3]	all	S2.2, S3.1, S4.1, S4.2	Not dependent on the BE typology	Always necessary in each case special consideration for hard targets or mass gathering events	employed technology
Users' involvement [S4.4]	all	S1.2, S2.2, S2.3, S4.1, S4.2, S4.3	Not dependent on the BE typology	Users should be trained to face disaster in all conditions	Financing informative campaign Types of guiding tools (e.g.: apps)

The classification used in this section highlights the complexities of relations among the physical and management-related elements within the BE, encompassing both outdoor and indoor spaces. This accomplishment stems from a meticulous consideration of robust regulatory frameworks and guidelines. The findings underscore the critical perspective that the BE and its occupants should not be construed merely as a backdrop for potential attacks but rather as integral components of the RMRSs themselves. Safety planners are advised to strategically coordinate two pivotal aspects: firstly, the design of the BE layout to facilitate spatial organization in regular usage, incorporating considerations such as standoff distances, and ensuring controlled areas and access under the purview of stakeholders; and secondly, BE-oriented interventions aimed at establishing secure perimeters and implementing constructive measures to safeguard building components, façades, and structures during emergency conditions, thereby mitigating the effects of terrorist acts [14, 23, 24, 32–34].

Table 2.6 Analysis of behavioural design factors for RMRS classes to consider in their design and evacuation simulation (N.A.: not assessed)

RMRS [code]	Interactions with behavioural issues	Possibility to be represented in crowd evacuation simulators
Design of the physical elements of the BE		
Safe perimeter [S1.1]	Barriers ought to be crafted with a thoughtful consideration of users' perceptions and behaviours during emergencies, such as evacuation, while maintaining a correlation with emergency layout and strategic planning	Geometry and obstacles can be represented in a virtual environment, studying the influence on the perpetrator and pedestrian evacuation dynamic
Secure envelope [S1.2]	N.A	Attack effects on the BE elements
BE layout		
Standoff [S2.1]	N.A	BE planimetric geometry
Sheltering [S2.2]	Their design should ensure the safety of users, addressing their essential needs in an emergency	Safe places are attractive for refuging
Emergency layout [S2.3]	Its design should consider the number of users and typologies to support the behaviour (literature or in simulation)	It constitutes input data for the setup of final conditions in simulation, influencing the evacuation paths to reach the defined safe areas
Access control and surveillance in the BE		
Access control [S3.1]	Aiming to discourage the perpetrators	It is an element/a set of elements influencing the pedestrian presence in the environment, representing input data in simulations
Security service [S3.2]	Aiming to discourage the perpetrators	Their incorporation into emergency scenarios enables the simulation of "intelligent" solutions, utilizing input data for the detection of emergencies and the management of evacuation
Illumination [S3.3]	Aiming to discourage the perpetrators	The degree of illumination affects the movement of individuals and influences the selection of specific paths, both in regular circumstances and during emergency evacuations
Safety and security management in the BE		
Security personnel [S4.1]	Aiming to discourage the perpetrators	It can be modelled as a source which modifies the pedestrian's evacuation

(continued)

Table 2.6 (continued)

RMRS [code]	Interactions with behavioural issues	Possibility to be represented in crowd evacuation simulators
First aid [S4.2]	Adequate to users' typologies and number	It can be expressed as a decrease in the number of victims and a directed movement of rescuers towards specific areas
Coordination [S4.3]	N.A	Simultaneous and coordinated employment of different countermeasures
Users' involvement [S4.4]	Instructions provided to users should align with their instinctive responses in hazardous situations	Capabilities of the users to perform proper safety behaviours

2.3 Factors Affecting the Terroristic Risk in the Outdoor Open Areas for the Most Recurrent Attack Typologies

As introduced in previous sections, the terroristic threat is a human-induced phenomenon, and its comprehension should consider three levels of elements:

- The perpetrator's will and decision capability.
- The BE features and uses.
- The user behaviour in evacuation and emergency processes.

These can be combined into two main issues which concern the main goals of the section in a BE-centred view, focusing on the assessment of its relations with both perpetrators and users.

In that sense, the discussion of the anthropic phenomenon and the current regulation and experiences framework at the international scale can support the interpretation and the parametrization of main features and properties related to the BE that should be considered in a risk assessment procedure.

Coherently with other risks, i.e. fire and earthquake, the risk assessment of a disaster usually considers tangible and intangible features related to the analysed elements (buildings, sub-components) in order to translate them into a final performance value towards a homogeneous system of elements to be compared, while users' behaviour has to be considered to understand and test mitigative strategies and solutions overcoming the risk dimensions towards resilient scenarios.

In this framework and in consequence of previous analyses presented in Sects. 2.1 and 2.2, a systematization of properties and elements affecting the risk assessment is discussed, in order to provide a limited set of elements to consider for a reduced and fast formulation of risk assessment.

Specifically, the attention is related to the OAs and the main attack types identified as efficient ones in Sect. 2.1, merging major results from the international experiences in mitigating the terroristic risks for mass gathering events and special/strategic targets. In the details of the summary presented in Table 2.7, nine recurrent keywords are identified to describe the terroristic risk which are discussed as follows, detailing the associated features:

- "TARGET" describes the type and inherent proneness of the place to be attacked. It is demonstrated by the phenomenological analysis of events in the BE and in OAs, highlighting the higher relevance in likelihood for CBEs B and D, consequently extending to F_B, F_D, and F for the events that occurred in relation to the outdoors (compare with Table 2.1). Specifically for OAs, the environmental significance of a location is contingent upon its inherent *likelihood* of being attacked, influenced by the notion of "soft target". Moreover, the size of the target does not preclude the symbolic importance of OAs. Even though the prior assessment of the terrorism phenomenon adopts a geographically independent analysis, the selection of an OA (one among other soft targets) should be tied to their *symbolic relevance* (i.e. religious, political, economic), which depends also on the presence of representative and symbolic buildings.

- "USES" is related to the impact "maximization" of violent acts, because the *use* of OAs and their structures assumes varying degrees of importance in terms of likelihood. Near the common uses of places, the "attractiveness" of squares/ streets or buildings facing OAs increases the potential proneness to perpetrators' choice, increasing locally the *touristic flows* [35]. Similarly, the presence of public buildings influences the use of the OAs, even if these are dependent on the opening times.

- "PREVENTION" considers the current significance of terrorism in urban environments, due to the fact that the extensive deployment of countermeasures or mitigative solutions can impact the potential likelihood of threats in OAs. This stems from the distinction between hard and soft targets (TP.3). Likewise, *preventive strategies* may vary based on weaponry and attack types aimed at achieving violent objectives (TP.4). In this context, the prevention encompasses both the existence of preventive measures in the urban BE (e.g., access control, robust barriers) and their efficacy against specific attack types (e.g., vehicular or armed assaults) [11, 35–38]. Thus, all mitigative urban physical elements, including geometric features of accesses, both within OAs and along their boundaries, participate in the discussion.

- "FORM/SHAPE" which discusses the morphological feature of the OAs and their relations with the assaults. This is strongly clear focusing on the attack typologies: for T2, mainly executed with cold arms or with guns, the perpetrator's violent act is "centralized" covering a circular area of interest; while for T3 the prevalent elongated features of places allow vehicles to reach higher speeds to pursue the act [12, 22, 36, 38, 39].

- "ACCESSIBILITY" related to the geometric dimensions of OAs while discussing their perimeter. The concept of accessibility is clearly stated in terms of the *physical permeability* of OAs as the ratio of physical geometries of accesses and the overall perimeter but also related to the urban regulations about *vehicular accessibility* (for the T3) [22, 36] or the topographic/human-induced conditions along the accesses (e.g., stairs, squatting) [40].

Table 2.7 Summary of recurrent keywords identified in the analysis of the collected background about the issues, including references and correlation with terrorism principles in Chap. 1, Sect. 1.1], classified by risk determinant type

Risk determ	Keyword	Terrorism principle	Contents	Refs.
HAZARD	TARGET	TP.3; TP.3.2	Inter-dependence and replacement of targets; soft target	[44]
		TP.2	Publicity impact is key to targeting	[44]
			Each EC has an inherent probability of being a target due to the relevance of being a soft target	[45]
			Symbolic value of the target; Presence of media	[45]
	USES	TP.1	Impact factor	[44]
			The potential high level of crowd of EC increases the likelihood of hazard	[45]
			People gathered in one place	[11]
			The level of alert could consider the attraction of places for tourists that can increase crowding	[35]
	PREVENTION	TP.3.1	Hard target	[44]
		TP.4	The characterization of terrorist weaponry	[45]
			Security personnel, the presence of the police force	[11]
			The introduction of countermeasures can prevent access to vehicles	[36, 38]
			Study strategies for controlling accesses	[35, 37]
VULNERABILITY	FORM/SHAPE		The presence of speed regulation elements limits the speed of vehicles along the street	[36, 38]

(continued)

Table 2.7 (continued)

Risk determ	Keyword	Terrorism principle	Contents	Refs.
	ACCESSIBILITY		The local topography of the place can preclude vehicle-borne threats	[22]
			Mitigative measures should be correctly designed to be effective	[36]
			Management of the vehicular traffic	[37, 46]
	OBSTACLES	TP.3.2	Soft target, not only as a place but also as a part of the place that allows high crowd levels (i.e. archaeological sites, stairs)	[44]
			Most of the "attractor" classes which have a high crowd level of people also outside the buildings (F_D–F_B) (i.e., Dehors)	[45]
			The presence of mobile or fixed obstacles being/as specific attractors for people (rendezvous, hangouts)	[37]
EXPOSURE	ATTACK TYPE		Inherent capacity of attack to maximize the effects	[45]
			Study different strategies related to possible attack types	[35, 37]
	CROWD	TP.1	The impact factor	[44]
			The high level of crowding influences the total number of victims	[45]

(continued)

Table 2.7 (continued)

Risk determ	Keyword	Terrorism principle	Contents	Refs.
			Check the variability of density in some parts of the places	[37]
	REACTION/ OBSTACLE		Use urban furniture or urban objects as protection during the attack	[41–43]
			Check the accesses and emergency paths and their capacity to be crossed during the evacuation	[37, 42, 43]
			Check along the accesses the presence of obstacles	[46]

- "OBSTACLES" recurs to describe all the OAs elements (within the area and along the frontier) that constitute temporal rendezvous for people. It's the case of bar-covered terraces, staircases, or greening that may increase locally the vulnerability of a place in terms of meeting points [37].
- "ATTACK TYPE" describes the relationships between the potential severity of the attack and the type of the attack itself. As demonstrated in previous sections, most of the mitigative strategies are classified coherently to the weapons or the means of the attack. On the other hand, the attack type itself constitutes the way to describe the severity of events when related to the OAs uses, as highlighted in the phenomenological analysis [35, 37].
- "CROWDING" is mostly related to the quantification of severity. In this case, the keyword is related to the maximum number of people to be involved in the events, considering the density of OAs and the associated facing buildings for their uses [37].
- "REACTION/OBSTACLE" describes the quality of OAs and its part in enhancing the responsiveness of users in the moment of the attack. Specifically, a first level of quality can be discussed focusing on the relationship between the ***physical objects/obstruction*** within the OAs and users. Here, their "protective" or "obstructive" potentialities can be considered [37, 41–43], following the main suggestions shared by some national guidelines to users: "hide" or "run".[3] The second level of discussion about the "reaction" refers to detailed ***countermeasures*** present within the OAs, assessed as effective for the attack types.

---

[3] The plans "Vigipirate" and "ACT—Action Counters Terrorism" [42, 43] summarize briefly the suggestion in the related French and English suggestions "s'échapper, se cacher, alerter et resister" and "hide, run and tell", promoted as smart guidelines for urban users involved in a terroristic acts.

The discussed keywords and their association with the risk determinants (hazard, vulnerability and exposure) offer the opportunity to parameterize the phenomenon in the OAs, combining specific boundary conditions. In fact, the recognized influence of the building uses in increasing or altering the proneness of events in squares and streets (CBEs F, B, D) and the attack types (T2 armed assault and T3 car bombing/car ramming) in Sect. 2.1 allow to limit the threat analysis towards a risk assessment of the phenomenological scenarios. As it is clear and fully argued in the literature and theory of risk assessment and management, the determination of simplified formulations for the analysis of scenarios can take advantage of collaborative methods, trying to overcome the limited knowledge about the issues while enhancing the single skills of other expert judgement.

References

1. Marone F (2013) La politica del terrorismo suicida. Rubbettino
2. Schwarzenbach A, LaFree G (2020) Political legitimacy and worldwide terrorist attacks, 1970–2017
3. The European Commission (2022) Security by design: protection of public spaces from terrorist attacks
4. Aven T (2012) On when to base event trees and fault trees on probability models and frequentist probabilities in quantitative risk assessments. Int J Perform Eng 8:311
5. Aven T, Renn O, Rosa EA (2011) On the ontological status of the concept of risk. Saf Sci 49:1074–1079. https://doi.org/10.1016/j.ssci.2011.04.015
6. Beňová P, Hošková-Mayerová Š, Navrátil J (2019) Terrorist attacks on selected soft targets. J Secur Sustain Issues 8:453–471. https://doi.org/10.9770/jssi.2019.8.3(13)
7. Bennett BT (2018) Understanding, assessing, and responding to terrorism: protecting critical infrastructure and personnel. John Wiley & Sons
8. Federal Emergency Management Agency (2009) Risk management series. Handbook for Rapid Visual Screening of Buildings to Evaluate Terrorism Risks (FEMA 455)
9. US department of Homeland Security (2018) Planning considerations: complex coordinated terrorist attacks
10. Home Office in partnership with the Department for Communities and Local Government (2012) Crowded places: the planning system and counter-terrorism
11. Kalvach Z (2016) Basics of soft targets protection–guidelines
12. Federal Emergency Management Agency (2011) Buildings and infrastructure protection series. Reference Manual to Mitigate Potential Terrorist Attacks Against Buildings (FEMA-426/BIPS-06), 2nd ed
13. Li Piani T (2018) Progettazione strutturale e funzione sociale dello spazio (qualc) vulnerabilità e soluzione al terrorismo urbano. Sicurezza, terrorismo e società - INTERNATIONAL JOURNAL - Italian Team for Security, erroristic Issues and Managing Emergencies (in italian; ISSN: 2421–4442) 7–15
14. Karlos V, Larcher M, Solomos G (2018) Review on soft target/public space protection guidance. Publications Office of the European Union, Luxemburg
15. Joint Counterterrorism Assessment Team (JCAT) (2018) Planning and preparedness can promote an effective response to a terrorist attack at open-access events
16. Bernardini G, Quagliarini E, D'Orazio M (2017) Grandi eventi e terrorismo: la progettazione consapevole della sicurezza delle persone. Antincendio 12 anno 69:12–28
17. Alnabulsi H, Drury J, Templeton A (2018) Predicting collective behaviour at the Hajj: place, space and the process of cooperation. Philos Trans Royal Soc B: Biol Sci 373:20170240. https://doi.org/10.1098/rstb.2017.0240

18. Templeton A, Drury J, Philippides A (2020) Placing large group relations into pedestrian dynamics: psychological crowds in counterflow. Collective Dyn 4:A23. https://doi.org/10.17815/CD.2019.23
19. Cuesta A, Abreu O, Balboa A, Alvear D (2019) A new approach to protect soft-targets from terrorist attacks. Saf Sci 120:877–885. https://doi.org/10.1016/j.ssci.2019.08.019
20. Laufs J, Borrion H, Bradford B (2020) Security and the smart city: a systematic review. Sustain Cities Soc 55:102023. https://doi.org/10.1016/j.scs.2020.102023
21. Jore SH (2019) The conceptual and scientific demarcation of security in contrast to safety. Euro J Secur Res 4:157–174. https://doi.org/10.1007/s41125-017-0021-9
22. NaCTSO - National Counter Terrorism Security Office (2017) Crowded places guidance. United Kingdom
23. Li S, Zhuang J, Shen S (2017) A three-stage evacuation decision-making and behavior model for the onset of an attack. Transp Res Part C: Emerg Technol 79:119–135. https://doi.org/10.1016/J.TRC.2017.03.008
24. Sommer M, Njå O, Lussand K (2017) Police officers' learning in relation to emergency management: a case study. Int J Disaster Risk Reduct 21:70–84. https://doi.org/10.1016/J.IJDRR.2016.11.003
25. Abreu O, Cuesta A, Balboa A, Alvear D (2019) On the use of stochastic simulations to explore the impact of human parameters on mass public shooting attacks. Saf Sci 120:941–949. https://doi.org/10.1016/j.ssci.2019.08.038
26. Zhu R, Lin J, Becerik-Gerber B, Li N (2020) Human-building-emergency interactions and their impact on emergency response performance: a review of the state of the art. Saf Sci 127:104691. https://doi.org/10.1016/j.ssci.2020.104691
27. Gayathri H, Aparna PM, Verma A (2017) A review of studies on understanding crowd dynamics in the context of crowd safety in mass religious gatherings. Int J Disaster Risk Reduct 25:82–91. https://doi.org/10.1016/j.ijdrr.2017.07.017
28. Ghazi NM, Abaas ZR (2019) Toward liveable commercial streets: a case study of Al-Karada inner street in Baghdad. Heliyon 5:e01652. https://doi.org/10.1016/j.heliyon.2019.e01652
29. Festag S (2017) Counterproductive (safety and security) strategies: the hazards of ignoring human behaviour. Process Saf Environ Prot 110:21–30. https://doi.org/10.1016/j.psep.2017.07.012
30. Liu Q (2020) A social force approach for the defensive strategy of security guards in a terrorist attack. Int J Disaster Risk Reduct 46:101605. https://doi.org/10.1016/j.ijdrr.2020.101605
31. Şahin C, Rokne J, Alhajj R (2019) Human behavior modeling for simulating evacuation of buildings during emergencies. Physica A 528:121432. https://doi.org/10.1016/j.physa.2019.121432
32. Gin JL, Stein JA, Heslin KC, Dobalian A (2014) Responding to risk: awareness and action after the September 11, 2001 terrorist attacks. Saf Sci 65:86–92. https://doi.org/10.1016/j.ssci.2014.01.001
33. Albores P, Shaw D (2008) Government preparedness: using simulation to prepare for a terrorist attack. Comput Oper Res 35:1924–1943. https://doi.org/10.1016/j.cor.2006.09.021
34. Wang J, Ni S, Shen S, Li S (2019) Empirical study of crowd dynamic in public gathering places during a terrorist attack event. Physica A 523:1–9. https://doi.org/10.1016/j.physa.2019.01.120
35. Ministère de l'Intérieur, Ministère de la Culture et de la Communication, Secrétariat Général de la Défense et de la Sécurité Nationale (2017) Gérer la sureté et la sécurité des événements et sites culturels
36. Centre for the protection of National Infrastructure (2014) Integrated Security. A Public Realm Design Guide for Hostile Vehicle Mitigation
37. Stadt Munster (2017) Public event safety. Guideline for creating a security of concept
38. (NaCTSO) NCTSO, Kingdom U, Infrastructure C for the P of N, Kingdom U (2012) Protecting Crowded Places: Design and Technical Issues
39. Federal Emergency Management Agency (2007) FEMA 430: Site and Urban Design for Security: Guidance Against Potential Terrorist Attacks

40. D'Amico A, Russo M, Bernabei L, et al (2022) A survey form for the characterization of the historical built environment prone to multi-risks. TEMA, Technologies Engineering Materials Architecture (e-ISSN 2421–4574) 8:77–88. https://doi.org/10.30682/tema0801b
41. Centre de Crise National TERRORISME ET EXTRÉMISME
42. National Counter Terrorism Security Office (2015) Stay Safe Film. In: 18 December 2015
43. Nationale SG de la D et de la S (2016) FAIRE FACE ENSEMBLE. VIGILANCE, PRÉVENTION ET PROTECTION FACE À LA MENACE TERRORISTE
44. Woo G (2015) Understanding the principles of terrorism risk modeling from Charlie Hebdo attack in Paris. Defence Against Terrorism Review-DATR 7:1–11
45. Cantatore E, Quagliarini E, Fatiguso F (2022) European cities prone to terrorist threats: phenomenological analysis of historical events towards risk matrices and an early parameterization of urban built environment outdoor areas. Sustainability 14:12301. https://doi.org/10.3390/su141912301
46. P. Säterhed, M. Hansson, J. Strandlund T, Nilsson, D. Nilsson, M. Locken AM (2011) Event Safety Guide. DanagårdLiTHO

Chapter 3
User Behaviour in Terrorist Acts to Model the Evacuation in Outdoor Open Areas

Abstract The resilience of the urban built environment to terrorist acts depends on the interactions among the physical scenario, the attackers, the hosted users, and the mitigation solutions (both structural and non-structural), when implemented. Outdoor Open Areas mainly show a high level of complexity in these terms, and thus, expert risk assessment methods to be applied in such contexts should be also supported by simulation-based approaches, which can be able to manage and describe these interactions in a holistic manner. The behavioural design approach can be used to evaluate the impact of different input conditions on final risk levels depending on the users' response to the terrorist act. In fact, this approach relies on the experimental-based modelling of user behaviours and individual vulnerability, and on the related simulation in emergency and evacuation scenarios. This Chapter hence traces bases for user behaviour modelling in terrorist acts.

Keywords Behavioural design · Simulation · Outdoor open areas · User behaviour in emergency · Evacuation · Terrorist acts

3.1 Understanding and Simulating User Behaviours in Terrorist Acts to Support Risk Assessment and Mitigation

As in different kinds of disasters (e.g., earthquakes, fires, floods) affecting the built environment (BE) [1–5], the behavioural response of the users can increase or decrease their risks in case of a terrorist act [6–10]. User behaviours thus represent an essential element to be considered in risk assessment and development of mitigation strategies [11–15], including both structural (mainly, physical interventions on the built environment) and non-structural (e.g., training and activities for risk perception, awareness and preparedness increase; emergency and evacuation planning, including the involvement of law enforcement agencies) measures [4]. Furthermore, the definition of behavioural patterns supports the definition and validation of terrorist act simulators and thus the possibility of adopting these tools

© The Author(s) 2025
G. Bernardini et al., *Terrorist Risk in Urban Outdoor Built Environment*,
SpringerBriefs in Architectural Design and Technology,
https://doi.org/10.1007/978-981-97-6965-0_3

for risk assessment and mitigation [14–17]. The behavioural design approach takes advantage of these knowledge and modelling standpoints and implies the analysis of experimental-based emergency behaviours of exposed users as the basic starting point for defining solutions against disasters [18]. Although this approach has been codified for other kinds of emergencies in the urban built environment, such as earthquakes [18], and it relies on the same perspective used in fire safety (e.g., according to the "Psychonomics" principles [19]) for buildings, recent works demonstrated their reliability also in the case of terrorist acts [8, 20]. Then, key performance indicators, based on the analysis of event impacts on the users and their behaviours, can quantitatively derive the risk levels in the built environment according to simulation results (see Chap. 4). The same modelling approaches could be used to assess risk in pre and post-retrofit scenarios, too.

In view of the above, this chapter first traces an overview of user behaviour in terrorist acts according to consolidated research (Sect. 3.2), also providing structured data on typical motion quantities (Sect. 3.3). Then, bases for simulation modelling are provided (Sect. 3.4) by using agent-based techniques, which can effectively represent the complex interactions among the outdoor Open Areas (OAs), the users and the perpetrators.

3.2 User Behaviour in Terrorist Acts

The analysis of users' behaviours can be mainly performed on videotapes of real-world events [8, 9, 15, 21]. Additional research methods involve the use of surveys (including those with survivors of real-world attacks, and those on hypothetical scenarios) [6, 7, 21–24], while recent efforts move towards virtual reality-based experiments [25, 26], although they are limited to indoor scenarios rather than to OA applications. Nevertheless, the analysis of real-world scenarios could be preferred since it represents a source with a low level of biases when it is performed by trained researchers, and thus, it can limit memory effects (e.g., in post-event interviews) and virtual spaces (e.g., realism, immersiveness, motion sickness) issues.

According to previous approaches [1, 2, 4, 5, 8, 21], user behaviours can be essentially organized in terms of evacuation phase (and thus of emergency and evacuation timeline), distinguishing three main phases. The pre-movement phase concerns the identification of possible emergency warnings and cues, and also includes preliminary tasks to decide if evacuating and the initial tasks (including evacuation direction identification). The motion phase represents the evacuation itself, and ends with the immediate post-evacuation phase, when users reach a safe area and try to re-organize tasks towards normality and reprise. Moreover, behaviours can be characterized in terms of the main issues composing the physical scenarios where the behaviours are performed (i.e., indoor/outdoor; presence of obstacles; presence of members of law enforcement agencies), as well as depending on the typology of attack (if statistically relevant or specific of given behaviour), and interaction elements. Each behaviour could also be classified as common with other kinds of emergencies or specific terrorist acts, and deliberately chosen or passively suffered. Finally, each behaviour can be associated with the probability of occurrence and situational frequency, which defines the possibility that they can be activated in emergency conditions depending on the aforementioned factors. Relying on structured results of previous works [8, 21], Tables 3.1, 3.2 and 3.3 organize these issues by respectively considering main behaviours in the pre-movement, evacuation motion, and immediate post-evacuation phase.

In general terms, although some situational frequencies could appear limited, the presence of related behaviours cannot be excluded, also in view of the restricted dimension of investigated samples. In this sense, the main scenario features defined in Tables 3.1, 3.2, and 3.3 can depict an increasing possibility that users can adopt specific behaviours. In this way, these tables also clearly report data for outdoor scenarios as the reference one in this work for OAs. Similarly, it is worth noting that such analyses were essentially consistent with "run and hide" procedures [27], and that fighting behaviours were not retrieved in the assessed conditions.

3.3 Summary of Main Motion Quantities in Terrorist Evacuation

Besides qualitative issues described in Sect. 3.1, as for other kinds of evacuation (e.g., earthquake, fire, flood) [2, 28–30], motion quantities in terrorist evacuation essentially concern pedestrian speed, and how pedestrian density, effects of the "modus operandi" of the attackers and specific typologies of scenarios could affect this speed. The need for experimentally-based data from real-world events is funda-mental to properly set up simulation models according to the effective quantities, rather than using generalized values (e.g., from general purposes databases). Never-theless, limited efforts seem to be made to this end, essentially in view of the lack of valuable data for the reliable analysis of user behaviours. In the following, most of the results have been collected by reference work (using videotapes of attacks all over Europe from 2004 to 2017) [8], while additional insights from other studies have been considered, too.

Considering free walking conditions (pedestrian density $\rho < 0.17$ persons/m^2), real-world scenarios (>600 records) point out that the instantaneous individual evac-uation speed V_i [m/s] ranges from 0.17 to 8.4 m/s (99th percentile of distribution), with a mean value of 3.32 m/s and a standard deviation of 1.93 m/s [8]. In this sense, values seem to be higher than those commonly noticed in general purpose and fire evacuation and adopted in related modelling (which essentially range from 1.2 to 1.5 m/s) [29, 30]. Normality for speed distribution is rejected, and data can be reliably described according to a Weibull distribution characterized by: mean $=$ 3.31 m/s, variance $=$ 3.76 m/s, scale $=$ 3.72 (standard error $= 0.08 =$, shape $= 1.77$ (standard error $= 0.06$), scale-shape covariance of parameters estimates < 0.007.

Table 3.1 User behaviours in the pre-movement phase according to structured results of previous works [8] (superscript *a*) and [21] (superscript *b*)

BEHAVIOURS: short description (issues of the behaviours which are: D = deliberately chosen; S = passively suffered)	Elements of interactions: *-main scenario features*	Situational frequency [%]
"PRO-SOCIAL" BEHAVIOURS*: Users engage in information searching and exchange for decision-making, i.e., activating or not the evacuation process and providing preliminary tasks for wayfinding (D)	Other users:	
	-general conditions	17[a]
	-near the attack area	20[a]
	-presence of safety/ security personnel	15[a]
RISK PERCEPTION AND EVACUATION DECISION DEPENDING ON SURROUNDING CONDITIONS*: The level of risk perceived by users changes with the presence of cues and triggers, and the evacuation procedure can be affected by the presence of sensible damages or effects of the attack (D). Moreover, the evacuation process can begin earlier for users who can directly observe triggers and cues of the attack with respect to others who are farther away from the attack area (S)	Sensible triggers and cues of the attack:	
	-overall effects	19[a] to 32[b]
	-near the attack area	25[a]
	-effective general modifications of the scenario due to the attack	19[a]
	-presence of safety/ security personnel	8[a]
	-arson	37[b]
	-bombing attack	60[b]
	-CBR attack	60[b]
	-melee attack	31 [b]
	-vehicle attack	50 [b]
	-shooting attack	47[b]
	-running crowd (in high-risk conditions)	75[b]
	-police action (in high-risk conditions)	42[b]
"CURIOSITY" EFFECTS*: Users can also decide not to evacuate, remaining close to their initial position, or moving more slowly in an attempt to "see what is happening", especially in case they are placed far from the event triggers and cues. Mainly, users can also take pictures or videos of the event through mobile devices (D)	Sensible triggers and cues of the attack, as well as other users who are evacuating or not:	
	-general conditions	42[a]
	-bombing attack	70[a]
	-outdoors	33[a]
	-presence of safety/ security personnel	48[a]

(continued)

Table 3.1 (continued)

BEHAVIOURS: short description (issues of the behaviours which are: D = deliberately chosen; S = passively suffered)	Elements of interactions: *-main scenario features*	Situational frequency [%]
	-effective general modifications of the scenario due to the attack	44[a]
	-far from the attack area	62[a]

*: the behaviour is noticed also in other kinds of emergencies (e.g., fires, earthquakes, floods)

Equation 3.1 describes the effects of ρ on V_i by adapting the factors of the equation of the fundamental diagram of pedestrian dynamics [31] depending on experimental values [8].

$$
V_i = \begin{cases} (2.50 - 0.72) * \left(1 - e^{-0.14*\left(\frac{1}{\rho} - \frac{1}{\rho_{crit}}\right)} \right) + 0.72 \text{ for } \rho \leq \rho_{crit} \\ k_L * (\rho - \rho_{crit}) + 0.72 \text{ for } \rho_{crit} < \rho \leq \rho_{stop} \end{cases} \tag{3.1}
$$

In Eq. 3.1, 2.50 m/s represents the free-flowing value of V_i, while 0.72 m/s refers to V_i for $\rho_{crit} = 2.67$ persons/m^2, that is for consolidated critical density values from real-world videotapes analysis.[1] $\rho_{stop} \geq 4$ persons/m^2 and considers the maximum values which can cause an evacuation stop [32]. Thus, while the V_i calculation $\rho \leq \rho_{crit}$ relies on experimental data, the one for $\rho_{crit} < \rho \leq \rho_{stop}$ has been defined by previous simulation works [20], theoretically hypothesizing a linear decreasing trend of V_i (where $k_L = $ -0.54 [m^3/(s•persons)]) due to the lack of consistent data on this part of the existence field of ρ [8].

Nevertheless, differences in V_i depending on the "modus operandi" of the attackers exist, also in view of the related effects and damage depending on the typology of terrorist act [11, 13, 33], as shown by Table 3.4.

These results from previous works [8] concern the average evacuation speed, which is hence elaborated by aggregating the instantaneous values V_i during the whole monitoring period. Table 3.4 also traces the average evacuation speed differences for outdoor and indoor scenarios. These data are combined regardless of the local pedestrian density, thus representing the average user behaviour in a significant part of the evacuation process. It is worth noting that data are calculated for a limited sample of users (<50 persons), and thus, they could be affected by dimensional biases and uncertainties.

---

[1] Data refers to the whole indoor and outdoor scenarios samples. In outdoors, Eq. 3.1 can conservatively assume that $V_i = 0.31$ for $\rho_{crit} = 2.67$ persons/m^2, with an exponential shaping correction factor of -0.19, according to the specific subsample of data [8].

Table 3.2 User behaviours in the evacuation motion phase according to structured results of previous works [8]

Behaviours: short description	Elements of interactions: *-main scenario features*	Situational frequency [%]
ATTRACTION TOWARDS SAFE AREAS*: Depending on the typology of the attack and physical scenario, try to move towards safe areas, generally distant from the event trigger or in protected zones (D)	Sensible triggers of the attack and physical scenarios:	
	-general conditions	63
	-far from the attack area	63
	-near the attack area	58
	-effective general modifications of the scenario due to the attack	68
	-presence of safety/security personnel	55
	-outdoors	58
	-by simply running far from the attack area towards the first available direction	28
"PRO-SOCIAL" BEHAVIOURS*: Social shared identity effects can support interactions during the motion phase, by supporting evacuation groups creation, information seeking and sharing (D). In addition, users' density alters the "collective" velocity of the group and thus the individual velocity (S). This behaviour includes the activation of specific responses depending on the surrounding conditions	Other users:	
	-general conditions	58
	-group ties between the users	32

(continued)

Table 3.2 (continued)

Behaviours: short description	Elements of interactions: *-main scenario features*	Situational frequency [%]
	-presence of more vulnerable users (e.g., hand assisted in evacuation, such as children, elderly, or disabled)	23
	-with respect to the activation of herding for path selection	41
	-presence of safety/security personnel	60
	-outdoors	52
	-bombing attack (as most relevant one)	78
	-effective general modifications of the scenario due to the attack	52
	-far from the attack area	62
REPULSIVE MECHANISMS TO AVOID PHYSICAL CONTACT*: users adapt their trajectory to locally avoid collisions with other users and obstacles (D)	Other users and obstacles:	
	-general conditions	17
	-outdoors	18
	-presence of safety/security personnel	19
	-presence of fixed obstacles	20
NOT KEEPING A "SAFETY DISTANCE" FROM FURNITURES*: Users allow physical contact with walls, fences, trees, indoor and urban furniture, chairs, railings, and movable obstacles since they are not perceived as unsafe for user movement. It also includes the possibility of climbing or knocking over such obstacles to optimize linear trajectories, limit directional changes or reduce waiting time along paths (D). The relevance of this behaviour could be also affected by users' density effects (S)	Movable obstacles:	

(continued)

Table 3.2 (continued)

Behaviours: short description	Elements of interactions: *-main scenario features*	Situational frequency [%]
	-general conditions	45
	-by climbing or knocking over them	20
	-effective general modifications of the scenario due to the attack	42
	-presence of safety/security personnel	30
	-near the attack area by climbing or knocking over them	28
	-high density of users (also over 1.33 persons/ m^2)	42
"SELFISH" AND COMPETITIVE BEHAVIOURS*: trampling or pushing behaviours are noticed in view of density increase and psychological pressure on the crowd while moving (D since the users activate this behaviour)	Other users and presence of triggers and cues of the attack, as well as attack typologies:	
	-general conditions	40
	-effective general modifications of the scenario due to the attack	41
	-near the attack area	45

(continued)

Table 3.2 (continued)

Behaviours: short description	Elements of interactions: *-main scenario features*	Situational frequency [%]
	-presence of safety/security personnel	18
	-vehicle attack (as the most relevant one)	58
INCREASED GUIDE EFFECT FOR PRESENCE OF RESCUERS*: leader–follower effects are noticed between safety/ security personnel (e.g., police officers, other first responders) and users. Users can take advantage of instructions from rescuers by mainly optimizing path selection and adopting protection behaviours (D)	Presence of safety/security personnel, as well as attack typologies:	
	-general conditions	22
	-outdoors	5
	-bombing attack (as the most relevant one)	41
	-near the attack area	45
AVOIDANCE OF EVACUATION PROCEDURE PERFORMING: Users can prefer adopting milling behaviours rather than evacuating, due to pro-social effects or curiosity effects (D)	Other users and presence of triggers and cues of the attack, as well as attack typologies:	
	-general conditions	34
	-far from the attack area	41
	-presence of safety/security personnel	31
	-vehicle attack (as the most relevant, i.e., for users not placed along the vehicle trajectory)	29

(continued)

Table 3.2 (continued)

Behaviours: short description	Elements of interactions: -main scenario features	Situational frequency [%]
	-armed assault (as the most dynamic in attackers' movement complexity)	20
COUNTERFLOW IN EVACUATION MOTION*: Groups of pedestrians may choose to go in opposing directions as a result of group behaviours or the identification of safe areas (D). This phenomenon can imply the group organization and shaping to reduce movement effort and collisions (S)	Other users and physical layout, as well as attack typologies:	
	-general conditions	28
	-presence of fixed obstacles	33
	-presence of safety/security personnel	15
	-vehicle attack (as the most relevant, due to the dynamic and rapid change of the attackers)	51
	-outdoors	30

*: the behaviour is noticed also in other kinds of emergencies (e.g., fires, earthquakes, floods)

3.4   Towards an Evacuation Model for Terrorist Acts Simulation in the Urban Outdoor Open Areas

As for other evacuation scenarios (e.g., fire, earthquake, general purposes) [4, 34, 35], an agent-based model (ABM) represents a suitable approach for terrorist acts simulation since it allows to consider the specific behaviours of the OA and its components, of the attackers and of the users, as well as their mutual interactions [36–40]. This approach can be easily combined with Cellular Automata (CA) techniques [41, 42], which divide the physical scenario (and thus the OA) into 2D cells in a quick but reliable manner. Due to the good balance between simulation outputs and execution timing, CA represents a useful technique to perform massive evacuation simulations [38, 43]. Moreover, ABM and CA have been combined by different simulation platforms, including open-source ones like NetLogo-based solutions [39], which have

Table 3.3 User behaviours in the immediate post-evacuation phase according to structured results of previous works [8]

Behaviours: short description	Elements of interactions: -*main scenario features*	Situational frequency [%]
Safe areas definition: Users typically stop the evacuation and gather as far away as possible from the attack area and damage due to the attack, where density conditions can also restore safety levels (D)	Sensible triggers of the attack, other users and physical scenarios, but noticed only outdoors:	
	-*general conditions*	26
	-*far from the attack area*	32
	-*effective general modifications of the scenario due to the attack*	30
	-*presence of safety/security personnel*	28
	-*evacuation conditions in low users' densities (up to about 0.30 persons/m²)*	92
	-*bombing attack (as the most relevant one)*	50
	-*considering the evacuation end for the influence of not immediate danger feelings or helplessness conditions (only this one includes indoor scenarios)*	16
"Pro-social" behaviours in post-evacuation*: In the immediate aftermath, as for other large-scale disasters (i.e., earthquakes, floods, typhons), users assist one another, especially considering more vulnerable and injured ones (D)	Other users, physical scenarios as well as attack typology	
	-*general conditions*	14
	-*outdoors*	17
	-*presence of safety/security personnel*	22
	-*armed assault (as the most relevant one)*	18
Attachment to things*: users try to move back and collect personal belongings, as for other large-scale disasters (i.e., earthquakes, floods, typhoons) (D)	Other users, physical scenario and attack typology:	
	-*general conditions*	17
	-*outdoors*	15
	-*presence of safety/security personnel*	21
	-*armed assault (as the most relevant one)*	20

*: The behaviour is noticed also in other kinds of emergencies (e.g., fires, earthquakes, floods)

Table 3.4 Average evacuation speed [m/s] depending on the typology of attack, and the type of scenario, in terms of minimum, mean and maximum values (approximated to 0.1). Data derived from [8]

Age typology (year range)	Minimum	Mean	Maximum
1-Typology of attack:			
(1.A) Bombing attacks	0.70	2.10	3.40
(1.B) Armed assaults with fire gun	1.80	2.50	3.20
(1.C) Attacks with a vehicle running into a target	2.00	3.20	5.00
(1.D) Other armed assault: spray	1.10	3.40	7.00
2-Scenario:			
(2.A) Outdoors	0.70	3.10	7.00
(2.B) Indoors	1.00	2.20	3.50

been widely used to perform evacuation simulations [36–38, 44]. Moreover, ABM-CA have been also selected, validated and applied by BE S^{2}ECURe in the context of terrorist acts simulation in OAs [20].

Figure 3.1 resumes the proposed ABM (intentional model), which is represented using the i* language representation [45]. The ABM is provided according to the main behaviours shown in Sect. 3.2. Each agent has its own resources to use/characterize itself, tasks to perform and goals to reach, while dependencies between them define the simulation rules inside each agent and consider their interactions with the other agents. In particular, the simulation issues concerning the user are organized according to the evacuation time, from the top (before the attack) to the bottom (evacuation completed).

In the following, the combined ABM-CA approach has been shown indeed according to these principles stressing issues concerning the users' exposure, vulnerability, and terrorist risk mitigation in the OAs developed within the project [11, 46]. Modelling issues are discussed in the following by involving the OA (Sect. 3.4.1), the attackers (Sect. 3.4.2), and the users (Sect. 3.4.3), and the resources, tasks, and goals of the ABM in Fig. 3.1 are highlighted in italics.

Figure 3.2 provides an overview of the CA approach from a spatial standpoint, thus representing the agents.

In the following, the combined ABM-CA approach has been shown indeed according to these principles stressing issues concerning the users' exposure, vulnerability, and terrorist risk mitigation in the OAs developed within the project [11, 46]. Modelling issues are discussed in the following by involving the OA (Sect. 3.4.1), the attackers (Sect. 3.4.2), and the users (Sect. 3.4.3), and the resources, tasks, and goals of the ABM in Fig. 3.1 are highlighted in italics.

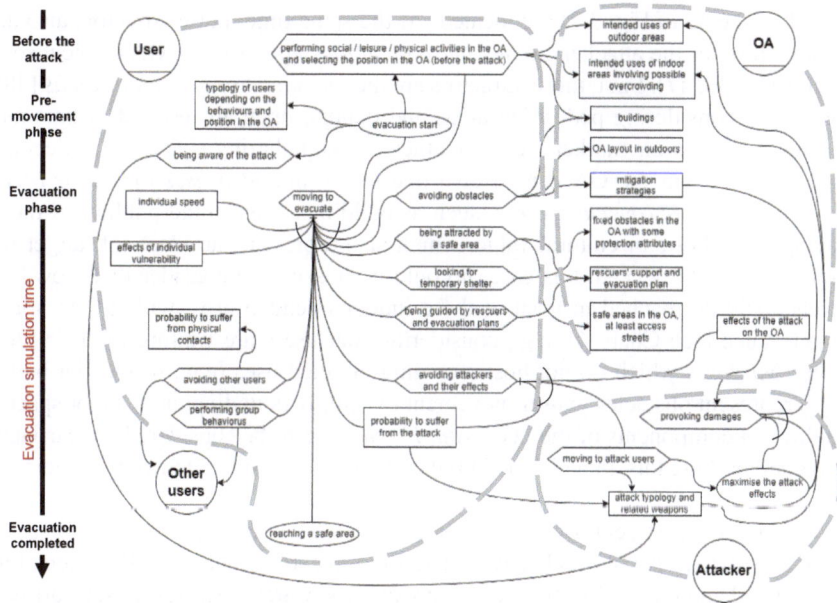

Fig. 3.1 Agent-based model for terrorist acts in the OA according to the i* representation. Each agents (circles) are characterized by specific resources (rectangles), tasks (hexagons), and goals (ellipses), placed under the agent's boundaries (dashed lines). General (arrows; "A depends on B" according to its direction) and contemporary (linked lines) dependencies are also traced

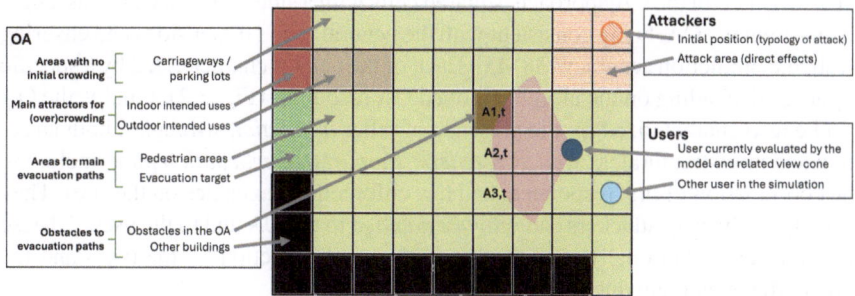

Fig. 3.2 Schematization of the considered model agents, divided into OA components, attackers and users), and main related typologies. A1,t, A2,t, and A3,t represent affordance values of some selectable cells by the users currently evaluated in the simulation at time *t*

3.4.1 Main Modelling Issues of the OA

The OA modelling focuses on outdoor spaces and areas surrounded by buildings. Before the attack, users can decide to *perform social/leisure/physical activities in the OA and select their position* outdoors where the main attractors for (over)crowding (see the scheme in Fig. 3.2) are located. At the same time, these attractors also act as attractors for the attackers to *maximize the attack effects* being OAs typical soft-targets [11, 33, 47]. First, these attractors are the *intended use of outdoor areas* [48], by mainly considering pedestrian areas and dehors, open-air terraces of bars and restaurants, open-air market areas, or other (mass)gathering spaces [46]. Second, (over)crowding in the OA can be also due to the *intended uses of indoor areas involving possible (over)crowding*, such as buildings open to the public and those having a symbolic or cultural value, which also represents an ideal soft target for the attackers [11, 47]. In this case, it could be possible to consider that crowding levels could be reached in front of these indoor intended uses, within their space of relevance (see Chap. 4), e.g., considering that users are waiting to enter [20]. Other *buildings* which do not host crowding-affected uses (e.g., residential build-ings) represent obstacles to user movement by simply bounding the outdoor spaces. Additional components of the *OA layout in outdoors* to be considered are carriage-ways and parking lots, in which no initial crowding is considered in view of their use by motor vehicles [46], and obstacles such as monuments, fountains, fences, shrubs and hedges, trees, street furniture, and other *fixed obstacles in the OA with protection attributes* such as passive and active barriers (engineered planters, wall barriers, low walls, fixed and retractable bollards, heavy objects, water obstacles, jersey barriers) [12]. In this sense, the OA also includes the *safe areas*, which represent the evacua-tion targets. They can be defined within the OA (e.g., physically surrounded by fixed protection obstacles), placed in the buildings surrounding the OA (i.e., according to "invacuation" strategies towards protected spaces[2]), or, at least, represented by the *access streets to the OA* in view of the need for the users to leave the attack-affected areas [11, 12]. In the CA approach, squared cells with a side of 50 cm can be assumed to represent the OA, being consistent with the general user's dimensions and ensuring a reliable prediction accuracy [38–43]. Each of the cells is characterized by a specific typology depending on the aforementioned OA resources (Fig. 3.2). Finally, the OA can be also characterized by the presence of other *mitigation strategies*, both struc-tural and non-structural, as well as *rescuers' support and evacuation plan* (including the coordination of first responders and law enforcement agencies on the site). They can impact the way attackers can *provoke damage* to the OA and to the users.[3] These resources can additionally alter the evacuation path selection by the users and the direct effects on them due to the attack.

---

[2] See e.g., https://www.gov.uk/government/publications/crowded-places-guidance/evacuation-inv acuation-lockdown-protected-spaces (last access: 01/12/2023). Although withdrawn, this guidance document provides a clear overview of evacuation versus "invacuation" strategies.

[3] For mitigation measures, please also compare with Chap. 4, Sect. 4.4.

3.4.2 Main Modelling Issues of the Attackers

The attackers are mainly aimed at *maximizing the attack effects* on the OA and the hosted users and thus their main task is *provoking damage* [38], while secondary goals could also concern the rapid escape from the OA after the attack, without being arrested by law enforcement.

The primary issue concerns the initial position of the attackers in the OA, which depends on the *intended uses of indoor and outdoor areas in the OA*. Considering the OA as an ideal soft target, the main attack attractors can essentially be identified by the most crowded areas and by the areas placed near targets with symbolic value, such as worship, public administration, and cultural buildings [33, 47, 49]. This initial position can be reached before the attack starts or during the attack itself.

Effects and damage, as well as the attackers' patterns, depend on the specific "modus operandi" and on the number and typology of involved attackers [11, 13, 33].

Bombing attacks could imply "static" effects in the simulation depending on the typology of the bomb and thus on the magnitude of the explosion and the radius of the effects, while direct movement of the attackers could be excluded from the simulation [27]. Similar issues can be linked to Chemical, Nuclear, Radiological, and Nuclear (CBRN) attacks [21, 33, 50], which can also involve a wider urban scenario apart from the OA.

Armed assaults are performed with different weapons [11], and they widely rely on a prey (the users)–predator (the attackers) model, in combination with the "shortest distance strategy", in which the predator essentially tries to prey on the closest users as the best attack preference [37, 44, 51–53]. Moreover, the effects on the user and the movement rules of attackers essentially depend on the selected weapon and on its "attack radius", which is a distance threshold for effective casualties. The overall approach essentially considers the following simulation steps:

1. The attackers, as predators, move and expand their search area until they find a user, as prey.
2. Once the prey is placed within their vision field, they will move chasing the user, preferably moving towards the nearest one.[4]
3. When the prey is placed within their vision field and within their attack radius, they will launch the attack and try to kill the user.
4. Then, the attackers will move towards a new prey, starting again from point 1 or 2 of the simulation steps.

Armed assaults with fire guns [11, 51, 54] are characterized by a wide distance threshold. Main behaviours essentially relate to the exploration of the OAs by single or multiple attackers, and their related possibility to remain in effective positions or move along effective paths for a significant time, shooting towards the users. Similarly, attacks with a cold weapon (e.g., knife, sword) can be performed by one or more attackers. The distance threshold for related casualties and the casualty rates

---

[4] As an alternative, they could move towards the more vulnerable users or towards specific targets in the crowd using the same logics of point 1 and 2.

depends on the typology of used cold weapons, but general values can range from 0.6 to 1 m radius [20, 42, 44, 51, 55]. This distance threshold can be associated with the probability that the attack can effectively provoke a casualty, in percentage terms [53]. The Terrorism Self-Aid Procedure (TSAP) probability threshold [%] can be hence associated with the users who suffer from the attacker's action [20], affecting the *probability to suffer from the attack* as the main resource in the ABM of Fig. 3.1 (see also Sect. 3.4.3). Thus, a casualty is provoked when the user-prey is placed inside the attacker-predator's threshold and if the user's TSAP is lower than a considered TSAP threshold. Nevertheless, this TSAP threshold can depend on the *weapon typology* and on the individual skills of a given user.

An attack with a vehicle running into a target is a typical outdoor attack in the OA [11, 33]. The target can be represented by the crowd (focused on a specific area or dispersed within the OA), a building (i.e., the vehicle moves towards the building façade or entrance), or a specific intended use placed outdoors, especially where (over)crowding levels or symbolic value are relevant. Besides the target, the vehicle driver can essentially adapt the vehicle trajectory to increase damage levels on the crowd. Although a significant lack in current literature is associated with the simulation of such type of attack, the proposed ABM model can be suitable to represent the related dynamics, by simply considering that the attacker corresponds to the vehicle itself and that the movement will be organized according to the possible microscopic trajectory of a vehicle. In this case, the attacker will mainly strike the users placed along the vehicle trajectory, while additional stampede effects could be simulated too [38] (compare with Sect. 3.4.3). The use of a distance threshold and a TSAP can be also considered for the attack with a vehicle. In particular, the distance threshold can be essentially considered equal to about half the vehicle width, thus considering users knocked down by the vehicle.

Attacks by unmanned vehicles/aircraft systems can essentially follow the same general rules of the attacks with a vehicle running into the target [11, 21]. Arson attacks can be essentially modelled according to fire-spreading dynamics [11].

Considering the other main Global Terrorism Database "modus operandi" [11, 56], it could be pointed out that unarmed assaults are specific typologies in which the crowd itself performs the attack as a whole, such as in the case of insurrections. In this case, the users (thus the crowd) and the attackers' dynamics are essentially overlapped towards a (soft) target. Similar issues are also related to barricade incidents, while facility/infrastructure attacks seem to be out of scope in this model with application to the OAs.

Finally, in risk assessment analysis, the ABM-CA model could also provide a "baseline" scenario condition referred to as the simple evacuation of the OA [52]. In this scenario, the attacker is not directly considered in the ABM and thus no effects of the attack on the users are generated. It hence allows to assess basic interactions between the users and the OA, regardless of the "modus operandi".

Table 3.5 Individual vulnerability by age typologies, including main motion features and ideal reduction of user speed to be applied to values calculated according to Eq. 3.1

Age typology (year range)	Motion features	V_i reduction [-]
Toddlers (0–4)	Assisted	0.53
Parents-assisted Children (5–14)	Assisted	0.87
Young Autonomous (15–19)	Autonomous	1.00
Adults A (20–69)	Autonomous	0.87
Elderlies E (70 +)	Autonomous or assisted	0.67

3.4.3 Main Modelling Issues of the Users

In all the attacks, the modelled emergency scenario implies the evacuation of the OA, since the attack is performed outdoors. Thus, it is considered that "users initially placed indoors do not need to participate in the evacuation process and can simply remain inside the buildings, where they are protected from the accident" [20]. Meanwhile, the initial position of users placed outdoors before the attack depends on the *social/leisure/physical activities performed in the OA* [46, 48], according to the intended uses of outdoor areas discussed in Sect. 3.4.1. Nevertheless, specific outdoor areas can attract specific *typologies of users depending on their behaviours*, and *individual vulnerabilities* and related features (e.g., by age, gender, and motion abilities) depending on their intended use. In this sense, at a broader level, users have to be also modelled to take into account their individual speed depending on age typologies [8, 32, 57], which implies an individual adaptation to the fundamental diagram shown in Eq. 3.1, by considering: (1) the ideal reduction factors on individual speed depending on age reported in Table 3.5; (2) the introduction of an individual random variation in speed (e.g., 0.7 m/s). According to the adoption of grid cells in the OA, the users' density ρ in Eq. 3.1 can be calculated according to the extended Moore neighbourhood approach [20, 44, 58]. In particular, the approach considers the cells that can be reached by the user i within 1 s of simulation time (as reaction time), at i's current speed. The density is calculated by excluding cells which are occupied by obstacles to evacuation paths. Moreover, the analysis could be limited to the cells placed within the users' view cone[5] and thus along the possible cells placed along the user movement direction, to consider the users' visual perception domain [41, 59, 60]. The use of this view cone can smooth the individual local trajectories by limiting sudden movements which are not experimentally noticed.

The *evacuation start* can be performed by users when they are *aware of the attack* [21]. The signal reception of information by the users led them to perform an initial about whether to evacuate. For instance, input data to this end could be correlated to huge sound levels, presence of smokes, individuation of suspected attackers or injured people, as well as surrounding crowd who has already started

---

[5] It corresponds to the horizontal field of view, and it is equal to 200° (https://bit.ly/3AYaCIY, accessed on 05/01/2024).

running and instructions by first responders. Thus, the start decision depends on the *attack typology and related weapons*, and it could be differentiated across the OA spaces, also depending on the position of the attack source, especially in case of attacks with cold weapons or vehicles running over the crowd. To consider these phenomena, individual pre-movement time [8, 32] can be modelled, thus including a delay between the attack starting and the evacuation start (e.g., depending on the distance from the attack area, recognition delays of the event). Nevertheless, two opposite but critical conditions could be identified, especially in dense crowd scenarios: (a) synchronous starting of user movement, which increases interactions among moving users; (b) activation of the evacuation start by distance from the attack source, as in the "Mexican waves" phenomenon [61], since moving users can impact those who are still waiting to start evacuating.

Users then start *moving to evacuate* the OA, by taking into account multiple tasks and resources as shown in Fig. 3.1, while *being attracted by a safe area*. The elements of reference for these behaviours essentially provoke attractive and repulsive phenomena in users' local and global paths, which are composed of the selection of different cells describing the OA. Thus, a dynamic floor field model "the willingness to walk" of a user placed in a certain cell towards one of the safe areas, according to a sort of affordance-based approach [40, 43]. Equation 3.2 provides the calculation of the affordance value $Aff_{c,t}$ [-] associated with a cell c of the grid, at time t, as proposed by previous research [20].

$$\text{Aff}_{c,t} = \alpha P_{i,c,t} + \beta F_c + \gamma R_{c,t} + \delta O_c. \tag{3.2}$$

$Aff_{c,t}$ dynamically changes over the simulation time depending on the composing factors, which are associated with related non-dimensional weights [40] (whose sum is equal to 1).[6] The probability that a user selects the cell c increases when $Aff_{c,t}$ increases. According to Eq. 3.2, these factors are:

- Dynamic, being time-dependent, to consider behaviours related to:
  Avoiding other users/performing group behaviours, by $P_{i,c,t}$ [-]. This factor considers the neighbouring pedestrian density with respect to the current position of user I, which is evaluated according to the abovementioned extended Moore neighbourhood approach [58]. $P_{i,c,t}$ is maximum where the pedestrian density is minimum, within the cells selected by the extended Moore neighbourhood approach. $P_{i,c,t}$ is associated with the weight α. When $\alpha \rightarrow 1$, *avoiding other users becomes* the prevalent behaviour in path selection. When $\alpha \rightarrow 0$, *performing group behaviours* are prevalent by the users placed in the same area, being the density negligible.
  Avoiding attackers and their effects, by $R_{c,t}$ [-]. This factor is introduced to consider the inclusion of a risk field for users' evacuation [42, 52, 55] and it

---

[6] Typical combinations of weights can be: $\alpha = 0$, $\beta = 1$, $\gamma = 0$, $\delta = 0$ for shortest path selection, e.g., in case of "baseline" scenarios (see Sect. 4.3.2) with no attackers; $\alpha = 0$, $\beta = 0.5$, $\gamma = 0.5$, $\delta = 0$ for attacks with weapons in which the attraction to safe areas has the same impact than running far from the attackers.

depends on the "modus operandi" of the attackers, according to Sect. 3.4.2. In particular, when no attacker ("baseline" scenario) is present, no effects are simulated and thus $R_{c,t} = 1$ for all the OA cells, and during the whole simulation time. In the other cases, $R_{c,t}$ increases with the distance from the attack area, but [37, 52, 53]: (1) for "static" attacks, e.g., bombing, $R_{c,t}$ is constant during the whole simulation time; (2) for attacks with different weapons, $R_{c,t}$ depends on the position of the attackers at the time t, and thus according to the defined prey (the evacuees)–predator (the attackers) model. In the case of more than one attack area, $R_{c,t}$ depends on the overlapping of the attack fields generated from each of the attack areas in the OA. $R_{c,t}$ is associated with the weight β. When $\beta \to 1$, the main users' goal in motion is to run far from obstacles.

- Static, being only layout-dependent, to consider behaviours related to:
    Being attracted by a safe area, by F_c [-]. This factor considers the distance from c to the closest *safe area in the OA*, thus overlapping the effects of different evacuation targets if present. In case no specific emergency plan is present, nor first responders tr to guide users and protect them from the attackers, it could be essentially considered that users try to move towards the OA access streets, far from the attackers, since these areas are perceived as safe [37, 42, 51–53, 55]. Different approaches can be used to define the calculation of this distance-based and wayfinding field, e.g., Dijkstra-based, A*, Priority Queue Flood Fill Algorithm [20, 21, 62–64]. The most distant cells are characterized by $F_c = 0$. The same approach could also take into account the activation of different safe areas over time to include behaviours related to *looking for temporary shelters* [21], according to the features of *fixed obstacles in the OA with protection attributes* as discussed in Sect. 3.4.1. In this case, their effectiveness, and thus the possibility to consider them as temporary shelters, depends on the specificities of the performed attack3. Moreover, the shielding effects of obstacles [21] or the visibility of safe areas [20] can locally alter the F_c values by respectively increasing or decreasing the considered distance and the wayfinding algorithm. F_c is associated with the weight γ. When $\gamma \to 1$, the main users essentially select the short evacuation path depending on the specific adopted algorithm.
    Avoiding obstacles, by O_c [-]. This factor considers the distance between c to the nearest obstacles to the evacuation path (see Fig. 3.2) if they are placed within the assumed interaction threshold of 3 m, which can cause modifications to the users' trajectory to avoid obstacles [60]. O_c is associated with the weight δ. When $\delta \to 0$, users allow for physical contact with obstacles.

Aff$_{c,t}$ varies from 0 to 1, since each composing factor in Eq. 3.2 is based on the normalization rules expressed by Equation 3.3, in which f_c is the value of a factor affecting *Aff*$_{c,t}$ considering c, and the subscripts *max* and *min* respectively describe maximum and minimum values among all the cells of the OA grid [20].

$$f_c = \left(f_{c,\max} - f_c\right) / \left(f_{c,\max} - f_{c,\min}\right). \tag{3.3}$$

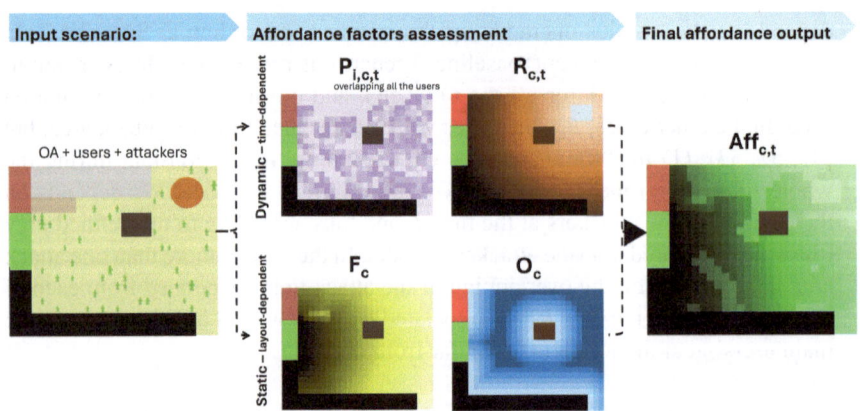

Fig. 3.3 Simulation workflow concerning the affordance calculation according to the adopted CA approach (see Eq. 3.2). In the panels representing CA maps for "affordance factors assessment" and "final affordance outputs", the colours of the cells range from lowest (light colours) to highest (dark colours) factor values

Figure 3.3 graphically shows the combination between the dynamic and static affordance factors described above, depending on the input scenario at a given time t, by tracing the related maps (the OA is divided into cells) and the overall $\text{Aff}_{c,t}$ map as the overlapping of them (in this case, all the weights are equal to 0.25 to overlap the related behavioural effects. It is worth noting that the factors in Eq. 3.2 could be integrated with attraction effects due to the presence of trained evacuation leaders [51, 65] (thus including attraction rules between pedestrians, rather than just repulsive phenomena as in $P_{i,c,t}$). In this sense, affiliative behaviours related to users' rescuing and support in motion (e.g., users trying to reach other injured users and then moving in close groups) [8] could be simulated according to the same criteria. Moreover, counterforce measures by law enforcement agencies can be also added to the model by considering, for instance, policemen fighting attackers and thus modifying $R_{c,t}$ and including them as new specific users within the model [15].

The surrounding conditions can also lead users to suffer from specific threats. Besides the *probability to suffer from the attack* (see Sect. 3.4.2), users can stop the evacuation process depending on *the probability to suffer from physical contact* and to be thus involved in falls [20, 32, 60]. Physical contact can appear evacuation in case of significant crowd density (>3 or 4 persons/m^2), of sudden reduction of the motion speed (deceleration > 0.3 g), of users moving in a counterflow, and of individual vulnerabilities (age or motion features related, e.g., elderly and assisted users could be more vulnerable to physical contacts). Probability thresholds to stop the evacuation can be then assigned to each user. In case the threshold is overcome, the *user* falls to the ground and should spend time rising up and restart moving. Previous works assigned a probability threshold equal to 5% and a random uniform distribution of fall time from 0 to 30 s [20, 32].

Previous works also tried to include "panic" effects within the terrorist act evacuation model [41], but these issues are not considered herein due to the poor validation by experimental-based data. Moreover, fighting behaviours are not modelled in Fig. 3.1 since they are limitedly noticed in real-world scenarios and law enforcement agencies' recommendations are essentially based on "run and hide" procedures (compare with Sect. 3.2).

From a simulation tool development, the CA model approach defined by Fig. 3.2 takes advantage of simulation time discretization [20, 32, 43]. The time step between two consecutive time t and $t + 1$ can be modelled depending on the maximum user speed, so as to represent the quickest evacuation process within the simulated agents [20]. Asynchronous update rules for user movement can be then considered, assuming: (1) a random selection in the users' simulation order at each step; (2) that each *user* can wait or move one cell per step by selecting the next one within the neighbouring ones placed along the movement direction and inside the view cone.

Finally, when *reaching a safe area*, the user exits from the simulation. Otherwise, the users can be removed from the model in case they suffer from the attack (being affected by casualties depending on TSAP, compare with Sect. 3.4.2) or when the maximum simulation time is reached.

References

1. Diakakis M (2020) Types of behavior of flood victims around floodwaters. Correlation with situational and demographic factors. Sustainability 12:4409. https://doi.org/10.3390/su1211 4409
2. Zhou J, Li S, Nie G et al (2018) Developing a database for pedestrians' earthquake emergency evacuation in indoor scenarios. PLoS ONE 13:e0197964. https://doi.org/10.1371/journal.pone. 0197964
3. Gwynne SMV, Boyce KE (2016) Engineering data. SFPE handbook of fire protection engineering. Springer, New York, New York, NY, pp 2429–2551
4. Bernardini G, Ferreira TM (2022) Emergency and evacuation management strategies in earthquakes: towards holistic and user-centered methodologies for their design and evaluation. In: Ferreira TM, Rodrigues H (eds) Seismic vulnerability assessment of civil engineering structures at multiple scales. Woodhead Publishing—Elsevier, pp 275–321
5. Wang Y, Kyriakidis M, Dang VN (2021) Incorporating human factors in emergency evacuation—an overview of behavioral factors and models. Int J Disas Risk Reduct 60:102254. https://doi.org/10.1016/j.ijdrr.2021.102254
6. Ludvigsen JAL, Millward P (2020) A security theater of dreams: supporters' responses to "safety" and "security" following the old trafford "fake bomb" evacuation. J Sport Soc Issues 44:3–21. https://doi.org/10.1177/0193723519881202
7. Bruyelle J-L, O'Neill C, El-Koursi E-M et al (2014) Improving the resilience of metro vehicle and passengers for an effective emergency response to terrorist attacks. Saf Sci 62:37–45. https://doi.org/10.1016/j.ssci.2013.07.022
8. Bernardini G, Quagliarini E (2021) Terrorist acts and pedestrians' behaviours: First insights on European contexts for evacuation modelling. Saf Sci 143:105405. https://doi.org/10.1016/j.ssci.2021.105405
9. Wang J, Ni S, Shen S, Li S (2019) Empirical study of crowd dynamic in public gathering places during a terrorist attack event. Physica A 523:1–9. https://doi.org/10.1016/j.physa.2019.01.120

10. Lovreglio R, Ngassa D-C, Rahouti A, et al (2021) Prototyping and testing a virtual reality counterterrorism serious game for active shooting. SSRN Electron J. https://doi.org/10.2139/ssrn.3995851

11. Quagliarini E, Fatiguso F, Lucesoli M et al (2021) Risk reduction strategies against terrorist acts in urban built environments: towards sustainable and human-centred challenges. Sustainability 13:901. https://doi.org/10.3390/su13020901

12. Federal Emergency Management Agency (2007) FEMA 430: site and urban design for security: guidance against potential terrorist attacks

13. US department of Homeland Security (2018) Planning Considerations: complex coordinated terrorist attacks

14. Bi L, Liu T, Liu Z et al (2023) Modeling heterogeneous behaviors with different strategies in a terrorist attack. Virtual Real Intell Hardware 5:351–365. https://doi.org/10.1016/j.vrih.2022.08.015

15. Liu Q (2020) A social force approach for the defensive strategy of security guards in a terrorist attack. Int J Disaster Risk Reduct 46:101605. https://doi.org/10.1016/j.ijdrr.2020.101605

16. Haghani M, Kuligowski E, Rajabifard A, Lentini P (2022) Fifty years of scholarly research on terrorism: Intellectual progression, structural composition, trends and knowledge gaps of the field. Int J Disaster Risk Reduct 68:102714. https://doi.org/10.1016/j.ijdrr.2021.102714

17. Liu H, Chen H, Hong R et al (2020) Mapping knowledge structure and research trends of emergency evacuation studies. Saf Sci 121:348–361. https://doi.org/10.1016/j.ssci.2019.09.020

18. Bernardini G, D'Orazio M, Quagliarini E (2016) Towards a "behavioural design" approach for seismic risk reduction strategies of buildings and their environment. Saf Sci 86:273–294. https://doi.org/10.1016/j.ssci.2016.03.010

19. Kobes M, Helsloot I, de Vries B, Post JG (2010) Building safety and human behaviour in fire: a literature review. Fire Saf J 45:1–11. https://doi.org/10.1016/j.firesaf.2009.08.005

20. Quagliarini E, Bernardini G, D'Orazio M (2023) How could increasing temperature scenarios alter the risk of terrorist acts in different historical squares? a simulation-based approach in typological Italian squares. Heritage 6:5151–5188. https://doi.org/10.3390/heritage6070274

21. Li S, Zhuang J, Shen S (2017) A three-stage evacuation decision-making and behavior model for the onset of an attack. Transp Res Part C: Emerg Technol 79:119–135. https://doi.org/10.1016/J.TRC.2017.03.008

22. Stevens G, Taylor M, Barr M et al (2009) Public perceptions of the threat of terrorist attack in Australia and anticipated compliance behaviours. Aust N Z J Public Health 33:339–346. https://doi.org/10.1111/j.1753-6405.2009.00405.x

23. Tahesh G, Abdulsattar H, Abou Zeid M, Chen C (2023) Risk perception and travel behavior under short-lead evacuation: post disaster analysis of 2020 Beirut Port Explosion. Int J Disaster Risk Reduct 89:103603. https://doi.org/10.1016/j.ijdrr.2023.103603

24. Tallach R, Einav S, Brohi K et al (2022) Learning from terrorist mass casualty incidents: a global survey. Br J Anaesth 128:e168–e179. https://doi.org/10.1016/j.bja.2021.10.003

25. Lovreglio R, Ngassa D, Rahouti A, Paes D (2022) Prototyping and testing a virtual reality counterterrorism serious game for active shooting. Int J Disaster Risk Reduct 82:103283. https://doi.org/10.1016/j.ijdrr.2022.103283

26. Liu R, Becerik-Gerber B, Lucas GM (2023) Effectiveness of VR-based training on improving occupants' response and preparedness for active shooter incidents. Saf Sci 164:106175. https://doi.org/10.1016/j.ssci.2023.106175

27. FEMA-426/BIPS-06 (2011) Reference Manual to Mitigate Potential Terrorist Attacks Against Buildings. FEMA-426/BIPS-06 Edition 2 510

28. Dias C, Rahman NA, Zaiter A (2021) Evacuation under flooded conditions: experimental investigation of the influence of water depth on walking behaviors. Int J Disaster Risk Reduct 58:102192. https://doi.org/10.1016/j.ijdrr.2021.102192

29. Hurley MJ, Gottuk DT, Hall JR et al (2016) SFPE handbook of fire protection engineering. Springer, New York, New York, NY

30. Shi L, Xie Q, Cheng X et al (2009) Developing a database for emergency evacuation model. Build Environ 44:1724–1729. https://doi.org/10.1016/j.buildenv.2008.11.008

31. Banerjee A, Maurya AK, Lämmel G (2018) Pedestrian flow characteristics and level of service on dissimilar facilities: A critical review. Collective Dyn 3:A17. https://doi.org/10.17815/CD.2018.17

32. van der Wal CN, Formolo D, Robinson MA, et al (2017) Simulating crowd evacuation with socio-cultural, cognitive, and emotional elements. Lecture Notes in Computer Science (including subseries Lecture Notes in Artificial Intelligence and Lecture Notes in Bioinformatics) 10480 LNCS:139–177. https://doi.org/10.1007/978-3-319-70647-4_11

33. The European Commission (2022) Security by design: protection of public spaces from terrorist attacks

34. Zheng X, Zhong T, Liu M (2009) Modeling crowd evacuation of a building based on seven methodological approaches. Build Environ 44:437–445. https://doi.org/10.1016/j.buildenv.2008.04.002

35. Kuligowski ED (2016) Computer evacuation models for buildings. SFPE handbook of fire protection engineering. Springer, New York, New York, NY, pp 2152–2180

36. Lu P, Li Y, Wen F, Chen D (2023) Agent-based modeling of mass shooting case with the counterforce of policemen. Complex Intell Syst. https://doi.org/10.1007/s40747-023-01003-9

37. Lu P, Wen F, Li Y, Chen D (2021) Multi-agent modeling of crowd dynamics under mass shooting cases. Chaos, Solitons Fractals 153:111513. https://doi.org/10.1016/j.chaos.2021.111513

38. Lu P, Zhang Z, Li M et al (2020) Agent-based modeling and simulations of terrorist attacks combined with stampedes. Knowl-Based Syst 205:106291. https://doi.org/10.1016/j.knosys.2020.106291

39. Banos A, Lang C, Marilleau N (2015) Agent-based spatial simulation with Netlogo. Elsevier

40. Liu R, Jiang D, Shi L (2016) Agent-based simulation of alternative classroom evacuation scenarios. Frontiers Architect Res 5:111–125. https://doi.org/10.1016/j.foar.2015.12.002

41. Song Y, Liu B, Li L, Liu J (2022) Modelling and simulation of crowd evacuation in terrorist attacks. Kybernetes. https://doi.org/10.1108/K-02-2022-0260

42. Chen C, Tong Y, Shi C, Qin W (2018) An extended model for describing pedestrian evacuation under the threat of artificial attack. Phys Lett A 382:2445–2454. https://doi.org/10.1016/j.physleta.2018.06.007

43. Li Y, Chen M, Dou Z et al (2019) A review of cellular automata models for crowd evacuation. Physica A 526:120752. https://doi.org/10.1016/j.physa.2019.03.117

44. Cao S, Qian J, Li X, Ni J (2022) Evacuation simulation considering the heterogeneity of pedestrian under terrorist attacks. Int J Disaster Risk Reduct 79:103203. https://doi.org/10.1016/j.ijdrr.2022.103203

45. Yu E (2009) Social Modeling and i *. In: Borgida A, Chaudhri V, Giorgini P, Yu E (eds) Conceptual Modeling: foundations and applications - Essays in Honor of John Mylopoulos. Springer, pp 99–111

46. Quagliarini E, Bernardini G, Romano G, D'Orazio M (2023) Users' vulnerability and exposure in public open spaces (squares): a novel way for accounting them in multi-risk scenarios. Cities 133:104160. https://doi.org/10.1016/j.cities.2022.104160

47. Lapkova D, Kotek L, Kralik L (2018) Soft Targets—Possibilities of Their Identification. In: Katalinic B (ed) Proceedings of the 29th DAAAM International Symposium. DAAAM International, Vienna, Austria, pp 0369–0377

48. Han S, Song D, Xu L et al (2022) Behaviour in public open spaces: a systematic review of studies with quantitative research methods. Build Environ 223:109444. https://doi.org/10.1016/j.buildenv.2022.109444

49. Cantatore E, Quagliarini E, Fatiguso F (2022) European cities prone to terrorist threats: phenomenological analysis of historical events towards risk matrices and an early parameterization of urban built environment outdoor areas. Sustainability 14:12301. https://doi.org/10.3390/su141912301

50. Bayram V, Yaman H (2024) A joint demand and supply management approach to large scale urban evacuation planning: evacuate or shelter-in-place, staging and dynamic resource allocation. Eur J Oper Res 313:171–191. https://doi.org/10.1016/j.ejor.2023.07.033

51. Arteaga C, Park J, Morris BT, Sharma S (2023) Effect of trained evacuation leaders on victims' safety during an active shooter incident. Saf Sci 158:105967. https://doi.org/10.1016/j.ssci.2022.105967
52. Yu H, Li X, Song W et al (2022) Pedestrian emergency evacuation model based on risk field under attack event. Physica A 606:128111. https://doi.org/10.1016/j.physa.2022.128111
53. Lu P, Li M, Zhang Z (2023) The crowd dynamics under terrorist attacks revealed by simulations of three-dimensional agents. Artific Intell Rev. https://doi.org/10.1007/s10462-023-10452-0
54. Zhu R, Lucas GM, Becerik-Gerber B et al (2022) The impact of security countermeasures on human behavior during active shooter incidents. Sci Rep 12:929. https://doi.org/10.1038/s41598-022-04922-8
55. Zhang F, Wu S, Song Z (2020) Crowd Evacuation during Slashing Terrorist Attack: A Multi-Agent Simulation Approach. In: Bae K-H, Feng B, Kim S, et al (eds) Proceedings of the 2020 Winter Simulation Conference. pp 206–217
56. National Consortium for the Study of Terrorism and Responses to Terrorism (START) (2019) Global terrorism database codebook: inclusion criteria and variables
57. Bosina E, Weidmann U (2017) Estimating pedestrian speed using aggregated literature data. Physica A 468:1–29. https://doi.org/10.1016/j.physa.2016.09.044
58. Hassanpour S, Rassafi AA (2021) Agent-based simulation for pedestrian evacuation behaviour using the affordance concept. KSCE J Civ Eng 25:1433–1445. https://doi.org/10.1007/s12205-021-0206-7
59. Xiao Q, Li J (2021) Evacuation model of emotional contagion crowd based on cellular automata. Discret Dyn Nat Soc 2021:1–18. https://doi.org/10.1155/2021/5549188
60. Lakoba TI, Kaup DJ, Finkelstein NM (2005) Modifications of the helbing-molnár-farkas-vicsek social force model for pedestrian evolution. SIMULATION 81:339–352. https://doi.org/10.1177/0037549705052772
61. Farkas I, Helbing D, Vicsek T (2002) Mexican waves in an excitable medium. Nature 419:131–132. https://doi.org/10.1038/419131a
62. Roan T-R, Haklay M, Ellul C (2011) Modified navigation algorithms in agent-based modelling for fire evacuation simulation. 11th International Conference on GeoComputation, London Session 2A:43–49
63. Chen L, Guo Z-L, Wang T et al (2023) An evacuation guidance model for heterogeneous populations in large-scale pedestrian facilities with multiple exits. Physica A 620:128740. https://doi.org/10.1016/j.physa.2023.128740
64. Syed Abdul Rahman SAF, Abdul Maulud KN, Pradhan B et al (2021) Impact of evacuation design parameter on users' evacuation time using a multi-agent simulation. Ain Shams Eng J 12:2355–2369. https://doi.org/10.1016/j.asej.2020.12.001
65. Kebir O, Nouaouri I, Rejab L, Ben Said L (2022) Simulating actors' behaviors within terrorist attacks scenarios based on a multi-agent system. In: Proceedings of the 12th International Defense and Homeland Security Simulation Workshop

Chapter 4
Measuring and Improving the Resilience of Outdoor Open Areas Against Terrorist Acts: A Behavioural Design Approach

Abstract The resilience of the urban outdoor built environment to terrorist acts depends on the interactions among the physical scenario, the attackers, the hosted users, and the mitigation solutions (both structural and non-structural), when implemented. Due to the complexity of the system, expert risk assessment methods should be also supported by simulation-based approaches. In this sense, this chapter first proposes a method to jointly consider hazard, vulnerability, and exposure in outdoor Open Areas (OAs) by then identifying possible emerging typologies and points of attack. Then, the behavioural design approach is used to evaluate the impact of different input conditions on final risk levels depending on the users' response to the terrorist act. In this sense, the quantification of user exposure and individual vulnerability is provided, since these parameters can vary over time and space, offering a complete view of input scenarios in case of terrorist act in the OAs. Then, the simulation of user behaviours in such defined emergency and evacuation scenarios can be performed thanks to experimental-based models. Key performance indicators (KPIs) are proposed herein to organize simulation results and quantitatively derive the risk levels in the built environment. Finally, regulation-based mitigation and protective strategies are identified, by considering implementation issues, but their effectiveness could be assessed by using the proposed behavioural-design-based methods taking advantage of simulation about the emergency and evacuation process.

Keywords Behavioural design · Risk assessment · Risk mitigation · Outdoor Open Areas · Key performance indicators

4.1 From Risk Scenarios to Risk Assessment and Mitigation in Outdoor Open Areas

As introduced in previous chapters, the risk assessment and mitigation of the terrorist threat in an outdoor Open Area (OA) are parts of a complex matter, since they are based on the joint analysis of features of the OA itself, perpetrator behaviours, and user behaviours and response to emergency conditions [1–4]. Moreover, supporting

© The Author(s) 2025
G. Bernardini et al., *Terrorist Risk in Urban Outdoor Built Environment*,
SpringerBriefs in Architectural Design and Technology,
https://doi.org/10.1007/978-981-97-6965-0_4

local authorities and their technicians to manage such issues is affected by the level of detail on available information and data, as well as of knowledge on the matter by safety designers. For this reason, methods should both pursue a qualitative and rapid standpoint, but also a quantitative and simulation-based approach, to ensure a complete understanding of possible risk scenarios and effects on the hosted crowd [5]. This chapter hence shows different methods for the creation of risk scenarios, assessment and mitigation effectiveness analysis, correlated to the elements that affect the phenomenology of terrorist acts.

In particular, a risk assessment method to provide possible attack points is defined depending on the effective features of the analysed OA (Sect. 4.2). Such method also supports the development of scenario creation concerning the desirability of the perpetrators in respect to the different specific areas and intended uses composing the OAs.

Then, in view of the dynamics of such a public open space [6], the time-dependent assessment of variations in the OAs use is discussed to evaluate users' exposure and vulnerability depending on the day and hour of the day in which a terrorist event can be performed (Sect. 4.3). In fact, OAs are typical soft-target for terrorist acts [7], and the probability, characteristics and modus operandi of a terrorist act, as well as its effects on the hosted users, strictly depend on the use of OAs spaces over time in view of dynamics at both the macro (urban) and micro (single OA) scales [6, 8–10]. Specific user-oriented key performance indicators (KPIs) are defined to this end.

Simulation-based approaches can exploit the results of such analysis to have a deep view of the emergency process, including the representation of evacuation behaviours and of the effects of the attack on the crowd, according to the model proposed in Chap. 3, Sect. 3.4. In particular, behavioural and simulation-based KPIs are herein defined and rules for the comparisons of scenarios (both in pre-retrofit conditions and in pre- versus post-retrofit conditions) are discussed (Sect. 4.4).

Finally, a rapid summary of Risk-Mitigation and Reduction Strategies (RMRSs) are also offered to support the decision-makers and designers' actions against terrorist acts in view of the user and OA-related features (Sect. 4.5).

4.2   Measure the Risk Assessment of Outdoor Open Areas to Provide Possible Attack Points in Real Case Study

The parametrization of the phenomenon characterizing terrorist threat in the OAs and the identification of boundary conditions can support the risk interpretation and resolution with smart approaches [11–13].

Such approaches require to be supported by qualitative and quantitative details related to the parameters involved [14, 15]. The parametrization process is already discussed in Sect. 2.3 and highlighted the major relationships among risk determinants (Hazard H, Vulnerability V, Exposure E) and prevalent features that describe OAs and the phenomenon. However, starting from previous study [16] the same parameters can be detailed towards the identification of logical and mathematic rules functional in solving the risk assessment formulation.

Specifically, as already discussed in [16], a system of indexes and parameters is setup in order to identify the global risk of real OAs. The risk formulation is based on three main assumptions at the basis of the structure in Fig. 4.1:

- The risk assessment R is structured in the three main determinants of risks, which are H, V and E (Eq. 4.1). H, V, and E are calculated as the combination of a limited set of indexes (i_n), as shown in Fig. 4.1 and Table 4.1, associated with a specific weight w which is coherently assigned according to expert judgment rules (Eqs. 4.2, 4.3 and 4.4). Moreover, R, H, V, and E are evaluated for single scenarios of attack types (*T-type*).
- Each of the indexes (i_n) is combined with one or more parameters (K summarized in Table 4.1) which describe qualitative and quantitative properties related to the given indexes. The K parameters, described in the following, are organized in order to have five classes of ranges, varying from 1 to 5, avoiding the risk equal to zero.

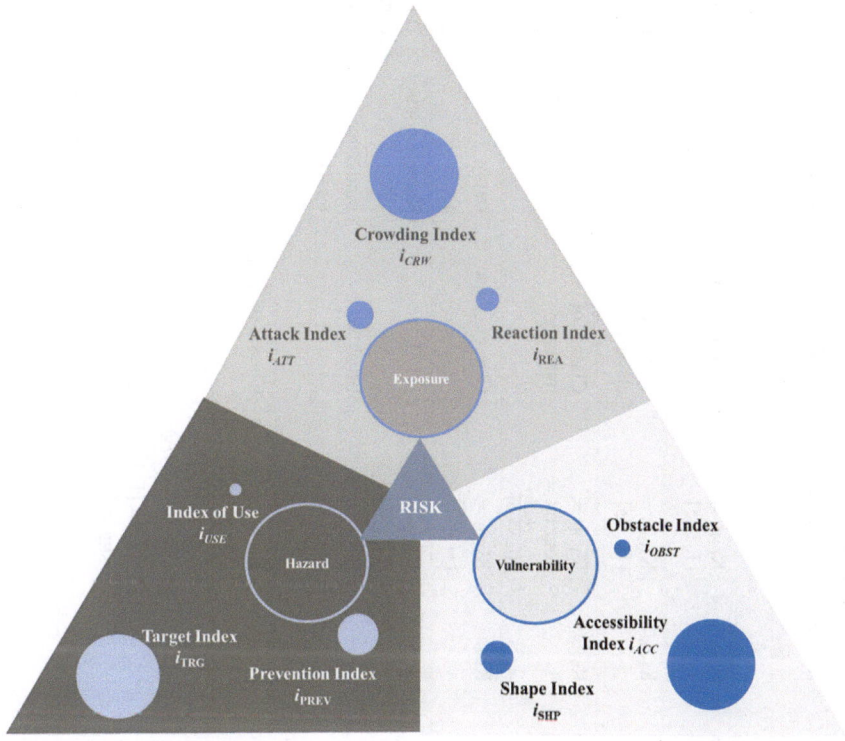

Fig. 4.1 Schematic structure of the risk assessment, organized in determinants and indexes

Table 4.1 Summary of indexes and k-parameters involved in the risk assessment formulation, detailing their equations, classification details, and range values

Index name i_n	K type	Equation	Classification details	Values				
				Remote	Unlikely	Possible	Likely	Very Likely
Hazard								
Target index i_{TRG}	K_{env}	$K_{ENV} = [1, …5]$	likelihood levels	1	2	3	4	5
	K_{symb}	$K_{symb} = [1, …5]$	symbolicity classes	negligible 1	low 2	medium 3	high 4	very high 5
Index of use i_{USE}	K_{TUR}	$K_{TUR} = Tour.Int = \frac{(n.arrivals)}{(n.inhab)}$	Classes of intensity	very low 1	low 2	medium 3	high 4	very high 5
	K_{use}	$K_{USE} = [1, …5]$	Classes of use	rarely 1	low 2	normal 3	high 4	very high 5
Prevention index i_{PREV}	K_{con}	$K_{CON} = \sum_{i=1}^{n} \frac{(Zi/Zeff)}{N.Access}$	Eff (T2)	Remote control	Direct/local control	Video Surveillance	Innovative systems	
			Eff (T3)	Innovative systems	Reinforced urban furniture	Barriers	Dissuasors	
Vulnerability								
Shape index i_{SHP}	K_{SHP}	$K_{SHP} = f_{EXT} × f_{SHP}$	Classes of f_{EXT}	0 < 2P/A < 0,02	0,02 ≤ 2P/A < 0,03	0,03 ≤ 2P/A < 0,06	0,06 ≤ 2P/A < 0,03	2P/A ≥ 0,09
		$f_{EXT} = [1, 5]$, $f_{SHP} = f(2P/A)$		1	2	3	4	5
		$f_{SHP} = [1, 1.5]$	Classes of f_{SHP}	Compact w/l ≥ 0.7		1.5 (T2)	1.0 (T3)	
		$f_{SHP} = f(w/l)$		elongated or very elongated f_{SHP} < 0.7		1.0 (T2)	1.5 (T3)	

(continued)

Table 4.1 (continued)

Index name i_n	K type	Equation	Classification details	Values				
Accessibility index i_{ACC}	K_{PER}	$K_{PER} = [1, 5]$ $r = \frac{\sum_{i=1}^{n}(A_{wi})}{2P}$	classes for r	0 < r < 0,05	0,05 < r < 0,1	0,1 < r < 0,2	0,2 < r < 0,3	r > 0,3
				1	2	3	4	5
	K_{ACC}	$K_{ACC} = \frac{\sum_{i=1}^{n}(Avi*facci)}{\sum_{i=1}^{n}Avi}$	$f_{acc} = [1,...,5]$	Not accessible	Limitedly	Moderately	Alternatively	Accessible
				1	2	3	4	5
Obstacle index i_{OBST}	$K_{OBST(V)}$	$K_{OBST} = \sum_{i=1}^{n} di * finf\, i$	$f\,inf = [1, 1.25, 1.5]$	No influence	Average increase	increasing		
		$d_i = Ai/Avi$		1	1.25	1.5		
Exposure								
Index of attack type i_{ATT}	K_{ATT}	$K_{ATT} = [4, 5]$	consequence levels for K_{ATT}	Minor	moderate	Medium	Major	Extreme
				1	2	3	4	5
Crowding index i_{CRW}	K_{CRW}	$K_{CRW} = [1, ...5]$	Occupancy classes for K_{CRW}	negligible	low	medium	high	Very high
				1	2	3	4	5
Index of attack reaction i_{REA}	$Kobst_{(E)}$	$K_{OBST(E)} = \sum_{i=1}^{n} di * finfi*$	f_{inf}	Decreasing	Average decreasing	not influential	average incremental	incremental
		$fshpobi$		0.5	0.75	1	1.25	1.5
		$fshpob$		negligible	low	medium	high	Very high
				1	2	3	4	5
	Kcm	$K_{CM} = Weff/Wi$ Wi = number of present contermeas	$W_{eff} = 3$	Alarm countermeasures	Evacuation countermeasures	Systems of physical interventions		

- The determinants of risk H, V, and E are evaluated for all the relevant Classes of Built Environment (compare with Chap. 2) present in the OAs, evaluated in the outdoor conditions, and thus for the square/street (F), and outside the public (F_B) and strategic/symbolic (F_D) buildings. In that sense, the identification of external area of public buildings takes advantage of the quantification process of the space of relevance (SoR) [17], as shown by Eq. 4.5. Here, the commercial extension of the public building ($A_{CommBuild}$ [m^2]) is related to the maximum density [persons/m^2] of buildings in indoors (C_B) and outdoors (C_{OUT}) coherently with fire safety regulations.[1]

$$R_{T-\text{type}(F.Fb,Fd)} = f\left(H_{T-\text{type}(F.Fb,Fd)}; V_{T-\text{type}(F.Fb,Fd)}; E_{T-\text{type}(F.Fb,Fd)}\right) \quad (4.1)$$

$$H_{T-\text{type}(F.Fb,Fd)} = ((i_{TRG} \times w_{TRG}) + (i_{Use} \times w_{Use}) + (i_{Prev} \times w_{Prev}))/w_{Tot} \quad (4.2)$$

$$V_{T-\text{type}(F.Fb,Fd)} = ((i_{SHP} \times w_{SHP}) + (i_{ACC} \times w_{ACC}) + (i_{Obst} \times w_{Obst}))/w_{Tot} \quad (4.3)$$

$$E_{T-\text{type}(F.Fb,Fd)} = ((i_{ATT} \times w_{ATT}) + (i_{Crw} \times w_{Crw}) + (i_{REA} \times w_{REA}))/w_{Tot} \quad (4.4)$$

$$A_{SoR}\left[m^2\right] = A_{CommBuild}\left[m^2\right] \times C_B[\text{persons}/m^2]/C_{OUT}[\text{persons}/m^2] \quad (4.5)$$

Given that rationale, Fig. 4.1 summarizes the qualitative and quantitative data structures of indexes and parameters involved, highlighting the major references.

As far as the significance of parameters, the main elements, properties, and details of values and ranges of the K parameters shown in Table 4.1 can be discussed as follows, in correlation with the related indexes shown in Fig. 4.1).

Hazard Indexes and K-Parameters

- The target index (i_{TRG}) assesses the symbolic significance of potential targets, taking into account political, religious, cultural and social factors. In that sense, the dimensions of relevance for standard uses and touristic attractiveness are translated in terms of K_{ENV}—which measures the statistical relevance of attacks for each environmental class (see level of likelihood in Chap. 2, Sect. 2.1), and K_{SYMB}—which quantifies the variation in symbolic significance of spaces. Both parameters help categorize the likelihood and symbolic importance of potential targets.
- The index of uses (i_{USE}) evaluates the attractiveness of places to perpetrators, independent of the number of people involved. For its description, K_{TUR} and K_{USE} are introduced. K_{TUR} reflects the inherent and potential representativeness of a place and its city, considering factors such as tourist influx and daily usage patterns.

---

[1] In this work, densities are correlated to Italian context in view of the application case study in Chap. 5, i.e. D.M. 03/08/2015 and National Ministerial Decree 19/8/1996; please compare also with Sect. 4.3.

K_{USE} describes the standard use of Open Areas and single structures, considering their inherent proneness to attacks based on daily usage patterns and conditions. These parameters aid in assessing the risk level associated with different urban spaces, providing insights into potential target selection by perpetrators.

- The prevention index (i_{PREV}) focuses on the presence of prevention strategies or solutions to mitigate terrorist attacks. The effectiveness of these measures depends on their relevance to the type of attack and the distinction between hard and soft targets. The effectiveness of strategies is already classified and discussed in Chap. 2, Sect. 2.2 by attack types (i.e., T2 and T3), and relates to remote control, direct/local control, video surveillance, and innovative systems such as face-detecting videos. In that sense, the quantitative parameter K_{CON} considers the presence and the number of protective systems for each possible access point to urban Open Areas, aiding in the assessment of their effectiveness in thwarting terrorist activities.

Vulnerability Indexes and K-Parameters

- The index of shape (i_{SHP}) focuses on the geometric configuration of OAs and its correlation with potential attack methods. K_{SHP}, representing the k-factor for this index, is determined by two factors: the extension of the OA (f_{EXT}) and the shape factor (f_{SHP}), which considers the relationship between width and length. Qualitatively, OAs are categorized as elongated or compact based on f$_{SHP}$ values. In fact, the vulnerability is influenced differently by OA morphology depending on the attack type; elongated spaces are more vulnerable to vehicle-based attacks (T3 with vehicle ramming), while compact spaces are vulnerable to centralized assaults (T2 with cold arms).

- The accessibility index (i_{ACC}) evaluates the ease of perpetrator access to OAs and it is described by means of K_{PER} and K_{ACC}. K_{PER} assesses the physical and geometric accessibility of the OA perimeter relying on the total width of OAs accesses (A_{vi} [m^2]) and the perimeter ($2P$ [m^2]); K_{ACC} considers the width of entrances and urban mobility features. In consequence of the latter, the accessibility levels vary between T2 and T3 attack types, with T2 being generally more accessible due to the significance of entrances, while T3 access is contingent on urban regulations and geometric constraints.

- The obstacle index (i_{OBST}) focuses on physical elements within OAs that may influence meeting and attractiveness in specific sub-areas. Elements such as urban furniture, terrain features, and gardens are evaluated in terms of their extension, relevance, and attractiveness influence. The obstacle parameter K_{OBST} is determined based on the ratio of obstacle extension (d_i) to the total obstacle surface and the associated attractiveness influence (f_{inf}).

Exposure Indexes and K-Parameters

- The attack index (i_{ATT}) assesses the potential level of people involved in attacks based on weapon types and attack methodologies. K_{ATT} quantifies the impact of weapon types as discussed in the phenomenological analysis in Chap. 2, Sect. 2. 1, using the classes of consequence levels.

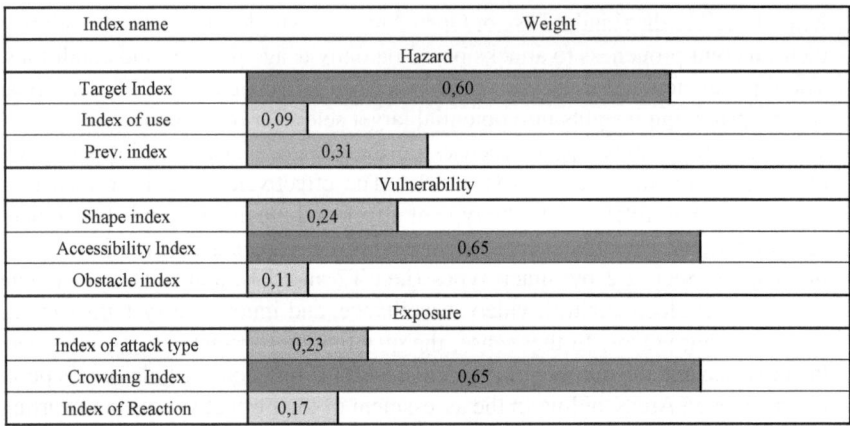

Index name	Weight		
Hazard			
Target Index		0,60	
Index of use	0,09		
Prev. index	0,31		
Vulnerability			
Shape index	0,24		
Accessibility Index		0,65	
Obstacle index	0,11		
Exposure			
Index of attack type	0,23		
Crowding Index		0,65	
Index of Reaction	0,17		

Fig. 4.2 Weights of indexes in the final formulation of risk determinants

- Crowd density influences exposure as well, represented by the crowd index (i_{CRW}), denoted as K_{CRW}, which considers the potential number of people involved in an attack scenario based on crowd density in Open Areas or surrounding public activities. The five ranges can be supported by the classification of uses for public spaces at the national level, when present.
- The index of the attack reaction (i_{REA}) evaluates the impact of physical elements in the environment on user reactions during an attack. It distinguishes between objects that can provide protective cover and those that hinder evacuation efforts. $K_{OBST(E)}$ quantifies the influence of obstacles and objects based on their extension, shape, and impact on protection or evacuation, coherently with the details discussed in literature [5]. Conversely, K_{CM} measures the positive effect of countermeasures on reducing the number of people involved in an attack. This considers strategies like alarm systems and evacuation plans tailored to different attack types (T2 and T3) (see Sect. 2.2).

All the presented K-parameters are valued following the rules of the participatory Delphi technique [18], in order to ensure the acceptability of relations among K-parameters and i_n indexes, as well as the formulation and ranging appropriateness of K-parameters.[2]

Finally, in order to solve the weighting of each index in the calculation of single determinants (Eqs. 4.2, 4.3, 4.4), an analytic hierarchy process (AHP) application has been processed[3] highlighting the higher relevance of three main indexes: Target, accessibility, and crowding indexes in each risk determinant (Fig. 4.2).

Even if the final aim of the formulation for the terrorist risk assessment presented in [16] is structured to support operative and comparative evaluation of real case

---

[2] The pool of participants is structured as a set of people already involved in the scientific studies and technical activities for the resilience and security of cities.

[3] The AHP methodology has been applied on the set of indexes by the same pool of participants.

Table 4.2 Details on classes of Risk determined for soft and hard targets considering the triad of values for each determinant and their combination

Target type	Level of danger	Class of risk
Soft target		
H [1, 2] ∧ E [1, 2] V [5]	all the combinations	Negligible
H [4, 5]	VxE = [1, 9]	Medium
V [1, 5]; E [1, 3]	VxE =]9, 15]	High
H [1, 2] V [1, 5]; E [3, 5]	VxE = [3, 6]	Low
	VxE =]6, 15]	Medium
	VxE =]15, 25]	High
H [3] V [1, 5]; E [1, 5]	VxE = [1, 4]	Low
	VxE =]4, 12]	Medium
	VxE =]12, 25]	High
Hard target		
H [4, 5] ∧ E [4, 5]	VxE = [4, 10]	Medium
V [1, 5]	VxE =]10, 25]	High

study, the same can be declined to determine possible attack points. Considering the structure of the formulation that provides a qualification of determinants for SoRs and street/square, and the geometric rules identified for the identification of SoRs in the OAs, the formulation can be focused on the single case study, providing a plan distribution of SoRs and risk properties. In the details, the structured formulation has determined a set of reduced bi-dimensional matrices which allow a brief discussion of the risk of OAs and their parts, where to variable condition of hazard proneness, levels of damages are determined for the setup of level of risks. Table 4.2 shows the details of such matrices, determined in [16], which became the way to qualify the OAs in all their parts (open space and SoRs), as specific target types.

4.3 Methods for Time-Dependent Assessment of Users-Related Factors

As for other kinds of emergencies affecting the urban built environment [19–21], the scenario creation in case of terrorist acts in OAs should consider the organization of data not only about OAs physical vulnerability/morphology and terrorist hazard (see Sect. 4.2), but also the user exposure and vulnerability. Recent works within the BE S²ECURe project[4] developed a joint approach for scenario creation based on the assessment of spatiotemporal variations of user-related factors depending on

---

[4] www.bes2ecure.net (last access: 16/10/2024).

the OAs characterization [6], so as to mainly derive inputs for emergency and evacuation simulation [22]. To pursue replicability and quick application, the proposed methodology essentially relies on:

Remote analysis via: (a) web mapping platforms such as Google Maps/Street view or Open Street Maps, to derive dimensions, typologies, intended uses, scheduling of areas in the OAs and the facing buildings, and to detect the presence of specific elements composing the OAs layout (including obstacles, street furniture); (b) national census databases, to determine the typologies of users depending on their age.

Standard occupant loads, such as those of fire safety codes, to determine a quick index of users' exposure by density [persons/m^2] depending on the intended use of the OAs and the facing buildings.

Nevertheless, the integration of specific GIS-based datasets and census data from local authorities can increase the accuracy of quick results. The whole methodology is shown in Fig. 4.3 and described above.

The first phase concerns the identification of intended uses placed outdoors and indoors and that can generate overcrowding in the OA, by also detecting the related surface, and of the user-related factors such as the main use behaviours, the quick occupant loads and the related temporalities.

Concerning the intended uses, the approach excludes residential areas since they essentially represent a sort of background level in users' exposure and vulnerability and have a limited impact on the terrorist act attraction due to negligible symbolic and strategical values [7, 23]. The approach hence considers outdoor and indoor areas characterized by users' gatherings (e.g., cinemas, sights, parks), public buildings, special buildings with symbolic value (e.g., worship places, museums) and hard

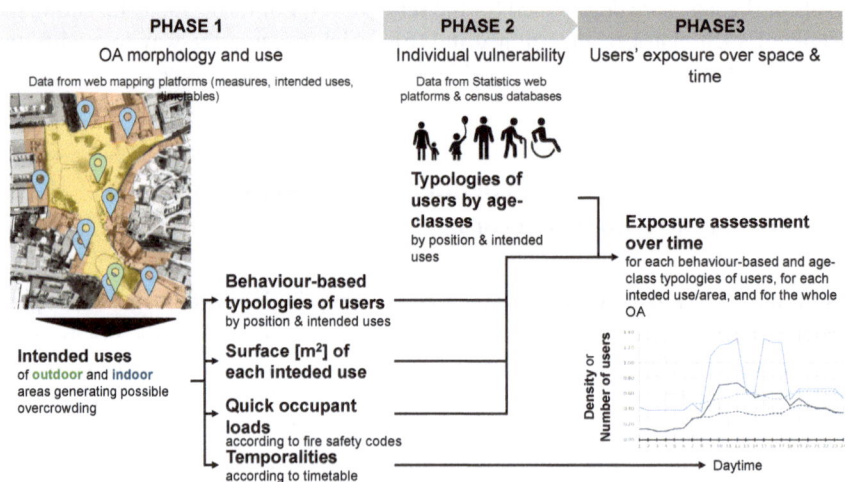

Fig. 4.3 General workflow for user-related factors assessment in the OAs according to BE S^{2}ECURe methodology [6]

targets. Indeed, areas not accessible to users such as fenced areas are excluded. Then, the available gross surface GS_i [m^2] is calculated for each selected indoor and outdoor area. Freeware web mapping tools can be used to measure the plan (or covered) gross surface of each area, e.g., by Calcmaps.[5] Then, for buildings, this surface is multiplied by the number of floors hosting the given intended use, e.g., by Google Street Maps.

Concerning use behaviours, different typologies of users are associated to the way they spend time in the intended uses, by mainly distinguishing behaviours between [8]: (a) only outdoor users (OO), who generally walk and move in the outdoors with a limited permanence times in the OA due to physical, social, and leisure activities, including sightseeing; (b) prevalent outdoor users (PO), who spend a long time walking in the outdoors or staying/sitting during social and leisure activities; (c) non-residents users (NR), who essentially populate buildings facing the OA and having a direct access to it, and could also contribute to the pedestrian volumes (moving towards or from the buildings) and gather in front of them while waiting to enter them. Access doors, gates and passages can be remotely identified by Google Street View.

Concerning quick occupant loads OL_i [persons/m^2] data from Italian fire safety codes [24] are herein adopted and combined with previous works assumptions to extend their applicability to both indoor and outdoor area [25, 26]. OL_i can be then arranged depending on specific data based on surveys. For example, OL_i can be substituted by using the number of seats instead for restaurants, cinemas and theatres, by the number of students and teachers for educational buildings, and by the number of workers for office buildings close to the public. Furthermore, OL_i can vary during the daytime depending on the scheduled activities of the intended uses. Then, temporalities for the considered areas can be derived according to timetables accessed via web search (e.g., opening times via Google Maps or websites of the specific activity open to the public), derived from national[6]/local regulations, or reasonable estimated by local habits [6, 25]. The occupant loads are applied to the timetable range while 0.00 persons/m^2 are considered out of the addressed timetable. The methodology also distinguishes between working days (as the most common and recurring conditions over the year) and holidays (Sundays and other national Holidays), because they can imply specific occupation variations both depending on timetable and use conditions. In addition, seasonal variations can be taken into account by revising OL_i, e.g., by increasing values for tourist destinations.

In view of the above, Table 4.3 resumes the selected intended uses categories associating typology of users and quick occupant loads, which are also determined in terms of related temporalities.

---

[5] https://www.calcmaps.com/map-area/ (last access: 28/11/2023).

[6] E.g.: https://www.mise.gov.it/index.php/it/mercato-e-consumatori/concorrenza-e-commercio/risposte-ai-quesiti/orari-di-apertura-e-chiusura (last access: 23/11/2023 – in Italian).

Table 4.3 Classification of intended uses by users' typologies and quick occupant loads according to fire safety code [24] and previous works [25] (i.e., for LOS, see [26])

Intended uses	Typologies of users (acronym)	Quick occupant loads OL_i [persons/m^2] and temporalities
Pedestrian areas (including sidewalks); green areas and parks accessible by users	Passersby as only outdoor users (OO)	Depending on the assumed level of service—LOS; some relevant classes can be: 0.00 (e.g., nighttime, from 1 to 6AM); 0.10 (LOS A, passersby's motion is totally free); 0.35 (LOS C, limit conditions for normal walking speed selection by passersby); 1.05 (LOS E, peak timings in passersby's presence in normal days with possible stoppages and interruptions of flows)
Dehors, open-air terraces of bars and restaurants	Prevalent outdoor users (PO)	≥ 0.4 for generic uses (in case of bars and restaurants: 0.7) during opening times
Open markets		≥ 0.4 during opening times
Outdoor mass gatherings areas (including temporary ones)		≥ 2.0 (up to 4.0) during mass gatherings; it can include relevant historical and cultural sites, and porticoes too
Educational buildings	Non-residents (NR)	0.4 during general lesson time (e.g., 8AM to 6PM for universities; 8AM to 2PM elsewhere) and 0.1 during office time (e.g., 2PM to 6PM) in working days; 0 during holidays
Hospitals, healthcare buildings, social welfare facilities		0.1 for ambulatory and 0.4 for visitors spaces during opening times; 0.1 for wards from 0 to 24 in both working days and holidays
Shops, other commercial buildings		0.4 during opening times
Bars, restaurants		0.7 during opening times
Government administrative buildings		0.4 for areas open to public and 0.1 for areas close to the public during opening times in working days; 0 during holidays
Worship places		0.7 at least during celebrations, for both working days and holidays; 0.4 or 0.7 in case of buildings with cultural and historical values attracting visitors (as for other cultural buildings and heritage)
Cinemas, theatres, auditorium and other similar recreational buildings		1.2 to 3.0, applied to the audience area/hall, during opening times in both working days and holidays
Cultural buildings and heritage, including museums and public libraries		0.4 or 0.2 (i.e., libraries) for general public areas, and 0.7 for visitors' gathering areas, during the opening times

(continued)

Table 4.3 (continued)

Intended uses	Typologies of users (acronym)	Quick occupant loads OL_i [persons/m^2] and temporalities
Transport stations		0.2, extended to the whole building area, during both working days and holidays
Office buildings, Factories and warehouses		0.4 for areas open to public, 0.1 for areas close to the public and 0.7 for workers/customers' gathering areas during opening times in working days; 0 during holidays
Accommodation facilities (e.g., hotels)		0.4, during both working days and holidays

The second phase concerns the collection of data on individual vulnerabilities, over space and time. Only outdoor users (OO), prevalent outdoor users, non-residents (NR) are users' typologies related to their position in the OAs, while additional users' typologies are related to their individual vulnerability due to age and gender. In fact, these factors can imply significant variations in the response to the emergency conditions in terms of pre-movement and evacuation behaviours, motion speed and susceptibility to direct/indirect damages from the attackers and the crowd phenomena [22, 27–29]. Age and gender data can be easily collected from local, regional and/or national census and statistics databases[30].[7] Municipalities-related distributions of population by age and gender can be considered valid for the urban areas, and thus for the OAs too, although refined on-site surveys can be then carried out at the microscale. According to a quick assessment approach, five age classes are assumed according to Chap. 3, Sect. 3.3, to represent motion issues and assistance needs in evacuation [27]. These classes are: toddlers T (0–4 years), parents-assisted children PA (5–14 years), young autonomous YA users (15–19 years), adult users AU (20–69 years), and Elderly users EU (70 + years). For each age class, the users' percentage by age class UP_{age} [%] is directly derived from population statistics databases. According to a quick assessment approach, UP_{age} is homogeneously considered for all the intended uses, except for educational buildings, where the age classes are referred to the typology of hosted students.

Then, temporalities of presence for users by age classes are associated to a presence coefficient at the given time t cp_t [-], which should be multiplied to UP_{age} and which varies from 0 (no user of the given age class for the considered intended use is present, thus $UP_{age} = 0\%$) to 1 (the number of users for the given age class is maximum, thus equal to UP_{age}). Such temporalities and thus cp_t can be assessed hourly. For OO, $cp_t = 0$ during nighttime, according to Table 4.3 insights, while for PO and NR, $cp_t = 1$ only during the opening times. Such data can be also refined by additionally considering the percentages [%] of male MU and female FU users, still according to the same databases. Moreover, additional user vulnerability factors can

---

[7] E.g., for the Italian scenario, National ISTAT annual reports on basic population statistics (i.e., percentage distribution by municipality) on age and gender for 2020: http://demo.istat.it/popres/index.php?anno=2020&lingua=ita (last access: 23/11/2023 – in Italian).

also relate to individual features regarding motion disabilities, since they can affect the motion and sensory abilities of the users [31]. Such kind of elements could be added in the proposed approach, by adding a specific presence coefficient and users' percentage by motion and sensory ability to Eq. 4.7. Nevertheless, related input data could be not available from a consolidated statistical perspective (i.e., using the same databases defined above), their collection could be time consuming, and data should be carefully collected and managed according to data protection authorities regulation since they could related to sensitive data, being related to individual health information.[8] Therefore, they are not considered in the proposed method since they are not easily managed according to quick and remote survey approaches.

The last phase concerns the time-dependent organization of data about users' exposure and individual vulnerability. From a wider perspective and considering a certain time t of the day (hour), the total number of users in the OA NU_t [persons] is calculated (Eq. 4.6) by summing the overall number of users by age-classes for each intended use i in the OA $NU_{age,i,t}$ [persons] (Eq. 4.7). NU_t dynamically varies from minimum (no user is present since intended uses are close to the public) to maximum (full opening of intended uses) conditions and describes the users' exposure without specifying individual vulnerability issues. In this sense, the total number of users by age in the OA $NU_{age,t}$ [persons] (Eq. 4.8) can describe the dynamics in individual vulnerability since it aggregates intended-based occupancy data by age classes.

Finally, the aggregation of $NU_{age,i,t}$ by intended uses having the same users' behaviours (OO, PO and NR) can be performed too, thus depicting vulnerability-related factors depending on the users' position and habits at the starting of the attack. In particular, in case the attack is performed in outdoors, as in considered in this work, the effective number of exposed users $NU_{t,exp}$ [persons] is equal to the number of users performing OO and PO behaviours, plus those preforming NR who are waiting to enter buildings.

$$NU_t = \sum_{age} \left(\sum_i NU_{age,i,t} \right) \tag{4.6}$$

$$NU_{age,i,t} = GS_i \cdot OL_i \cdot cp_t \cdot UP_{age} \tag{4.7}$$

$$NU_{age,t} = \sum_i NU_{age,i,t} \tag{4.8}$$

In view of the above, Table 4.4 summarizes the KPIs on time-dependent users' exposure and vulnerability. All the KPIs can be used to compare different risk scenarios within the same OA, and between several OAs, since they concern values which are normalized or expressed in reference to the OA surface data. Some KPIs can be also used to provide input data for emergency simulations.

---

[8] https://commission.europa.eu/law/law-topic/data-protection/reform/rules-business-and-organisations/legal-grounds-processing-data/sensitive-data/what-personal-data-considered-sensitive_en (last access: 26/02/2024).

Table 4.4 Key performance indicators proposed for describing time-dependent dynamics in users' exposure and vulnerability, derived from the approach of the BE S^{2}ECURe project [6]

KPI —*acronym* [unit of measure]	*Calculation*—use	Range
Overall users' outdoor density in outdoor at a given time t - UOd_t [persons/m^2]*	*Ratio between NU$_t$ and the overall OA surface*—quickly comparing different time conditions in terms of users' density for the same OA and for different OAs (higher the density, higher the exposure)	0 to 3 persons/m^2 (reasonable condition for overcrowding)
Users' normalized number at a given time t - NUn_t [-]	*Ratio between NU$_t$ and the maximum daily value of NU$_t$ for the given condition*—scaling the conditions at the given time of the day to the maximum reference conditions in terms of users hosted in the built environment. To be applied within the same OA, or in different OAs if normalizing by the maximum NU$_t$ value in all the considered OAs	0 (excluded) to 1 (included, as the crowded time of the considered period)
Impact of an event in the OA on the whole population at a given time t - IE_t [-]	*Ratio between the sum of NU$_t$ referred to only outdoor intended uses (that is, the OA itself), and NU$_t$*—assessing the possible impact of risks in outdoors by excluding users who can look for shelter indoors at the starting of the attack. TO be applied within the same OA and in different OAs	0 (minimum risk since all the users are indoors) to 1 (maximum risk since all the users are outdoors)

(continued)

Table 4.4 (continued)

KPI —acronym [unit of measure]	Calculation—use	Range
Percentage of users by position at the given time t - OOp_t, POp_t, NRp_t [%]*	Ratio between the users by their position and NU_t —assessing the vulnerability of users depending on their initial position among different scenarios in the same OA and in several OAs. OO and PO can be directly exposed to the attack in the OA all over the time	0 to 100
Percentage of users by age and gender given time t - Tp_t, PCp_t, YAp_t, AUp_t, EUp_t, MUp_t, EUp_t, FUp_t [%]*	Directly from statistics databases or as the ratio between $NU_{age,t}$ and NU_t —assessing the individual vulnerability in comparable terms among different scenarios in the same OA and in several OAs. Critical values can be retrieved if values about T, PC, and EU are maximized	0 to 100

*: the KPIs can be used also for simulation scenario creation

Furthermore, the KPIs proposed in Table 4.4 are organized over the time t, still using the hourly sampling mentioned above, and they can be also assessed by separately referring to working days, holidays, and exceptional use conditions (e.g., fairs, exhibitions, concerts, other one-off events and mass gatherings). Furthermore, KPIs can vary depending on the season or the day of the week, depending on the specific OA use conditions. KPIs statistics on maximum values and average (for normal data distribution) or median (in case of non-normal data distributions) values can be calculated regardless of time, to respectively provide a quick description of critical and recurring OA conditions.

4.4 Simulation-Based Indicators

Emergency and evacuation simulations can be performed through the model defined in Chap. 3, Sect. 3.4, by using input data about the OA morphology and layout, the position, quantity and quality of exposed users (Sect. 4.3), and the quantity and quality of the points of attack (Sect. 4.2). Due to the stochastic effects related to the users' behaviours within the simulation models (i.e., initial user distributions, individual speed calculation, path selection and motion loops), a significant number of runs repetition (≥ 10) has to be performed for each scenario, and the general convergence indicators shown in Table 4.5 should be evaluated [22, 32, 33]. These indicators can be analysed to evaluate if the number of consecutive runs is enough to provide statistically-reliable simulation outputs. Thresholds for each indicator varies depending on the given acceptance criteria, but general works remarks that evacuation time and related standard deviation should be at least $\leq 5 \div 10\%$ to ensure confident preliminary analysis.

Besides, convergence analysis, the statistical-based analysis of simulation results has to be performed also to derive KPIs for risk-assessment purposes [22, 34]. The first level of aggregation of data concerns the definition of the evacuation curve for the given scenario, since it traces the overall effects of evacuation interactions between users, the OA and its components, the attackers and their effects on the users and the OA. If the normality of simulation results could not be confirmed, the median evacuation curve, expressing the median number of users reaching a safe area (for the OA, one of the access streets) over time, should be considered. In fact, median values refer to the 50th percentile of distributions and they seem to be robust enough to trace results being not easily affected by extreme values in distributions [35]. In additional quartile-based curves (e.g., 5th, 25th, 75th and 95th) can be calculated. In particular, the curves referring the 5th and 95th percentile of users arrived at a safe are over time can trace the reasonable limits for the effectiveness of the simulated scenario, excluding behavioural outliers [33].

Table 4.5 Main simulation convergence indicators according to literature works [32, 33]

Convergence indicator—*acronym* [unit of measure]	Calculation	Meaning
Average Total Evacuation Time – $TET_{av,j}$ [s]	$TET_{av,j}$ is equal to the average maximum evacuation time TET_j of each j-th run in the given set of runs	The indicator expresses the time needed by the last user to complete the evacuation. The difference between two consecutive $TET_{av,j}$ should tend to 0
Average Evacuation time at the 95% of arrived evacuees – $T95_{av,j}$ [s]	$T95_{av,j}$ is equal to the average maximum evacuation time $T95_j$ of each j-th run in the given set of runs	The indicator excludes possible behavioural outliers in users' evacuation due to model uncertainties and subtitlies, e.g., unfavourable conditions in initial position of the user within the OA, evacuation path choice, interaction with other users and individual speed
Standard Deviation of total evacuation time – SD [s]	Standard deviation of the total evacuation time for the given set of runs	The indicator is consistent assuming the normal distribution of evacuation times. The value can be calculated also for T95
Euclidean Relative Difference – *ERD* [-]	$ERD = \dfrac{\lVert \vec{x} - \vec{y} \rVert}{\lVert \vec{y} \rVert}$	Similarity of angle two curves exists if ERD tends to 0
Secant Cosine – *SC* [-]	$SC = \dfrac{<\vec{x}, \vec{y}>}{\lVert \vec{x} \rVert \lVert \vec{y} \rVert}$	Similarity of shape between two curves, considering their first derivative, exists if SC tends to 1
Euclidean Projection Coefficient – *EPC* [-]	$EPC = \dfrac{<\vec{x}, \vec{y}>}{\lVert y \rVert^2}$	Similarity in the translation of the points that compose the curve, thus describing a sort of scale factor, exists if EPC tends to 1
Difference between the graphic Areas Under the Curves – *DAUC* [%]	$DAUC = \dfrac{\int \vec{x} - \int \vec{y}}{\int \vec{y}} \bullet 100$	Similarity in the "rapidity" of the evacuation process over time, by considering the whole area under the curves, exists if DAUC tends to 0%

$\vec{x}$ and $\vec{y}$ represent the average curves of two sets of consecutive runs (e.g., considering 10 runs, $\vec{y}$ refers to runs 1 and 9 and is the reference curve, while $\vec{x}$ refers to average curves from all the runs and is the curve to be checked)

While these curves trace a time-based overview of the evacuation process, the KPIs listed in Table 4.6 summarize the overall risk conditions of a given terrorist act scenario in the OA. These indicators have been developed within the BE S^{2}ECURe project to trace main behavioural issues in terrorist act evacuation [22], and to be consistent with previous works also concerning other kinds of emergencies, such as general purposes, fire and earthquake [36–40]. To ensure the KPIs robustness [35], they take advantage of median values from simulation results on the set of simulation runs are considered as for the evacuation curve.

Moreover, the KPIs are normalized to make them ranging from 0 (minimum risk) to 1 (maximum risk). Therefore, they can compare different input scenarios on the same effects scale. In that sense, they can both compare several conditions related to the current scenario of the analysed OA, e.g., as in pre-retrofit conditions, by varying the simulation input factors related to the attack typology, the points of attack, the users' exposure and vulnerability depending on the time of the day. Similarly, they can be used to compare pre-retrofit scenarios with post-retrofit scenarios implementing specific RMRSs, given that these RMRSs can modify the OA layout, the effects of the attack, the spatiotemporal distribution of the users, and also the user behaviours in emergency and evacuation. To this end, simulation models should be adapted to represent possible specific behaviours apart from those defined in Chap. 3, Sect. 3.4.

The comparison between KPIs can be then performed in absolute terms, as the difference between the KPIs, since all of them range from 0 to 1, so as to derive how specific conditions *scen* can impact the KPI levels in respect of a given reference scenario *ref*. Nevertheless, percentage variation of a given KPI *PV* [%] can be calculated according to Eq. 4.9, which is based on previous works on behavioural-based design [41]. *PV*-based assessment can better stress the final KPI levels in respect of the original one. Indeed, both absolute differences of the KPIs < 0 and *PV* $< 0\%$ imply an increase in the users' safety considering the KPIs in Table 4.6.

$$PV_{scen,ref} = \frac{KPI_{scen} - KPI_{ref}}{KPI_{ref}} \cdot 100[\%] \qquad (4.9)$$

Table 4.6 Key performance indicators for evacuation risk assessment in case of terrorist acts in the OA, based on simulation results, and derived from the approach of the BE S²ECURe project [22]

Simulation KPI—*acronym* [unit of measure]	Calculation	Meaning
Normalized evacuation time at the 95th percentile of arrived users—*TN95* [-]	$TN95 = \frac{T95_{av,j}}{T_{max}}$, where T_{max} [s] is the maximum simulation time (when the simulation ends, compare with evacuation model variables in Chap. 3, Sect. 3.4)	It expresses the time during which users can be still exposed to the attackers in the OA, since some of the are still placed inside it, by excluding outliers (compare with $T95_{av,j}$ in Table 4.5)
Normalized flows at the 95th percentile of arrived users— *FN95* [-]	$FN95 =$ $\max\left(0.1 - \frac{(F95/\sum l_s)}{1.5 persons/s/m}\right)$ where l_s [m] is the width of the access street to the OA	It expresses the speediness of the evacuation process since it relies on the slope of the curve (represented by the users' flow in persons/s). 1.5 persons/s/m is the normalization reference by representing the maximum specific users' flow from previous literature works [42]
Normalized number of physical contacts among the users—*PN* [-]	$PN = \frac{(PC_{T95}/T95)}{PC_{max}}$ where PC_{T95} [events] represents the effective (simulation-based) number of physical contacts and PC_{max} [events/s] is the maximum number of physical contacts, equal to $5\% NU_{t,exp}$ per second [events/s]	It assessed crowd dynamics and interferences by comparing the effective and maximum physical contacts per second (5% of exposed users as reference probability threshold to stop the evacuation [29]). Dividing PC_{T95} by $T95$ allows deriving other indicators that can be compared in different scenarios and for different $T95$. When PN increases, effects of overcrowding and interactions with OA obstacles are more relevant
Casualty ratio—*CR* [-]	ratio between the number of user casualties due to the attackers and $NU_{t,exp}$	It expresses the impact of the attackers' action on the crowd, and thus depends on the attackers' strategy. At least, *CR* is equal to 0 in case no attacker is present (e.g., a "false alarm" scenario)
Not-arrived users' ratio—*NA* [-]	ratio between the number of users who did not complete the evacuation during the simulation time and $NU_{t,exp}$	*NA* includes the effects due both to casualties and any other user who does not leave the OA (e.g., because they prefer gathering in areas inside the OA; compare with modelling details in Chap. 3, Sect. 3.4). In this sense, it also depends on the OA morphology and the attackers' strategy

4.5 Mitigation and Preventive Strategies Towards Effectiveness and Outdoor Open Areas Compatibility

The mitigation of risk and the prevention of violent acts are related to the qualification of risk itself, the identification of intrinsic vulnerabilities and the full knowledge of the phenomenon. In that sense, the application of expeditious formulation for risk assessment combined with the setting up of possible critical scenarios in the real OAs involving users' behaviour and the use of fast performance indicators allow the analysis of the identified scenarios in an as-built conditions. The reduction of risk and the improvement of resilience of urban place and users can be achieved through a comprehensive and effective system of strategies, which comprehend physical and technical solutions. If the effectiveness of such strategies can be solved by the standards and regulations, also coherently with the attack types, the compatibility of strategies requires to be declined in terms of compatibility of solutions with the real OAs. As already highlighted in discussed in the theories of the "design of security" in British counterterrorism activities [7, 43, 44], the transformation of the physical places may affect the integrity of a real place and the security perception of its users. Considering the relevance of cultural and symbolic places in the proneness to a terroristic attack [45], the resolution of compatibility became pivotal when applied in cultural or historic places. In that sense, all the RMRS strategies identified for their classification (discussed in Sect. 2.2), a system of datasheets is setup, linking to each potential physical element involved in the strategies, technical solutions, and their possible levels of physical or aesthetical compatibility. The solutions are derived from the analysis of the current regulation about countermeasures of terrorist threat at the international level and properly highlighted in the following Table 4.7.

Table 4.7 Classification and description of regulation-based RMRSs (according to classes provided in Chap. 2), including discussion on their implementation details and forecasted efficacy against main terrorist act typologies

Class	Design of the physical elements of the BE [S1]			
Sub-class	ANTI-RAM URBAN FURNITURE [AF] 1/2			
SUB-CATEGORY	[AF_1] TREES	[AF_2] NOGO BARRIER	[AF_3] BLOCK	[AF_4] FLOWERPOT
Type of functioning	Passive	Passive	Passive	Passive
Description	As system, prevention or limitation of the passage of vehicles (T3). System of trees can support the temporary protection to cold arms (T2)	Prevention or limitation of the passage of vehicles (T3). As System can support the temporary protection to cold arms (T2)	Prevention or limitation of the passage of vehicles (T3). Its use can be combined with other systems	Prevention or limitation of passage of vehicles (T3), also combined with other systems. Associated to higher dimension can provide temporary protection to cold arms (T2)
Installation type	Permanent	Permanent	Temporary/ permanent	Temporary/ permanent
Presence of foundation	Natural, Superficial or deep	Rested on the ground/ pavement	Rested on the ground/ pavement	Shallow foundation
Anti-ram	Yes	Yes	Yes	Yes
Certificate/ test	N.a	N.a	Vehicle 7,5 t ≤ 80 km/h	Vehicle 7,5 t ≤ 80 km/h
Source	[46]	[46]	[44]	[44]
Main materials	Greenery	Metal	Stone	Stone, cement
Accessibility	Pedestrians, Bicycles, Wheelchairs	Pedestrians, bicycles, wheelchairs	Pedestrians, bicycles, wheelchairs	Pedestrians, bicycles, wheelchairs
Integrability	Building' distance	As Artwork	Materials and shapes	Materials and shapes
Possible interferences	Urban surface network	Any	Any	Any
Efficacy (T2)	Medium	Medium	Not relevant	Medium
Efficacy (T3)	Medium	High	High	Medium
OAs compatibility	High	High	H igh	High
Class	Design of the physical elements of the BE [S1]			
Sub-Class	ANTI-RAM URBAN FURNITURE [AF] 2/2			

(continued)

Table 4.7 (continued)

Class	Design of the physical elements of the BE [S1]			
SUB-CATEGORY	[AF_5] ENGINEERED PLANTER	[AF_6] HEAVY OBJECTS	[AF_7] BENCH	[AF_8] SEATS
Functioning type	Passive	Passive	Passive	Passive
Description	Preventing or limiting the passage of vehicles (T3), also in combination with other systems. Extending dimensions, it can provide temporary protection to cold arms (T2)	Heavy objects (monuments, sculptures) for preventing or limiting the passage of vehicles (T3). Extended dimensions can provide temporary protection to cold arms (T2)	Useful for preventing or limiting the passage of vehicles (T3). Its use can be combined with other systems	Useful for preventing or limiting the passage of vehicles (T3). Its use can be combined with other systems
Installation type	Permanent	Temporary/ permanent	Permanent	Permanent
Foundation	Variable deep	Rested on the ground/pavement	Shallow foundation	Shallow foundation
Anti-ram	Yes	Yes	Yes	Yes
Certificate/test	Variable	n.a	Vehicle 7,5 t ≤ 80 km/h	n.a
Source	[46]	[46]	[44]	[44]
Materials	Cement	stoNe, cement, metal	Wood, stone, cement	Stone, cement
Accessibility	Pedestrians, bicycles, wheelchairs	Pedestrians, bicycles, wheelchairs	Pedestrians, bicycles, wheelchairs	Pedestrians, bicycles, wheelchairs
Integrability	Materials	As artwork	Materials and shapes	Materials and shapes
Interferences	Urban surface network	Any	Any	Any
Efficacy (T2)	Medium	High	Not relevant	Not relevant
Efficacy (T3)	High	Medium	Medium	High
OAs compatibility	Medium	High	High	High
Class of measure	Design of the physical elements of the BE [S1]			
Category	ANTI-RAM BARRIER [AB] 1/2			

(continued)

Table 4.7 (continued)

Class of measure	Design of the physical elements of the BE [S1]			
SUB-CATEGORY	[AB_1] MOBILE WEDGE BARRIER	[AB_2] ROTATING WEDGE	[AB_3] RISING WEDGE BARRIERS	[AB_4] FIXED JERSEY BARRIER
Functioning type	Active	Active	Active	Passive
Description	Retractable mobile barrier for limiting the passage of vehicles (T3)	Retractable fixed barrier for limiting the passage of vehicles (T3)	Retractable fixed barrier for limiting the passage of vehicles (T3)	Fixed barrier for limiting the passage of vehicles (T3). Extending dimensions, it can provide temporary protection to cold arms (T2)
Installation type	Permanent	Permanent	Permanent	Permanent
Foundation	Absent	Deep foundation	Shallow foundation	Shallow foundation
Anti-ram	Not	Yes	Yes	Yes
Certificate/test	n.a	Vehicle 7,5 t ≤ 80 km/h	Vehicle 7,5 t ≤ 80 km/h	Vehicle 5 t ≤ 80 km/h
Source	[46]	[46]	[44]	[44]
Materials	Iron	Iron	Iron	Reinforced concrete
Accessibility	Controlled (vehicles)	Controlled (vehicles)	Controlled (vehicles)	Denied (vehicles)
Integrability	Not possible	Not possible	Retractable	Not possible
interferences	Any	Urban surface network	Any	Urban surface network
Efficacy (T2)	Not relevant	Not relevant	Not relevant	Medium
Efficacy (T3)	Medium	High	High	High
OAs compatibility	Low	Low	High	Low
Measure Class	Design of the physical elements of the BE [S1]			
Category	ANTI-RAM BARRIER [AB] 2/2			
SUB-CATEGORY	[AB_5] MOBILE JERSEY BARRIER	[AB_6] MODULAR BARRIER	[AB_8] DROP-ARM CRASH BEAM	[AB_8] ROD
Functioning type	Passive	Passive	Active	Active

(continued)

Table 4.7   (continued)

Measure Class	Design of the physical elements of the BE [S1]			
Description	Mobile barrier useful for limiting the passage of vehicles (T3). Extending dimensions, it can provide temporary protection to cold arms (T2)	Mobile device useful for limiting the passage of vehicles (T3)	Mobile device useful for limiting the passage of vehicles (T3)	Fixed device useful for limiting the passage of vehicles (T3)
Installation type	Temporary	Temporary	Permanent	Temporary/permanent
Foundation	Absent	Absent	Absent	Rested on the ground/pavement
Anti-ram	Not	Yes	Yes	Yes
Certificate/test	n.a	n.a	n.a	Vehicle 7 t ≤ 80 km/h
Source	[46]	[46]	[44]	[44]
Materials	Reinforced concrete	Iron	Reinforced concrete	Reinforced concrete, iron
Accessibility	Denied (vehicles)	Controlled (vehicles)	Pedestrians, bicycles, wheelchairs	Controlled (vehicles)
Integrability	Not possible	Not possible	Not possible	Not possible
interferences	Any	Any	Any	Any
Efficacy (T2)	Low	Not relevant	Not relevant	Not relevant
Efficacy (T3)	Medium	High	High	High
OAs compatibility	Low	Low	Low	Low
Measure Class	Design of the physical elements of the BE [S1]			
Category	BOLLARDS [BO] 1/2			
SUB-CATEGORY	[BO_1] FIXED	[BO_2] DEEP AND FIXED		[BO_3] SHALLOW
Functioning type	Passive	Passive		Passive
Description	Road device that simulates the anti-ram effect. reduce the probability of attack occurring with vehicles (T3)	Useful for preventing or limiting the passage of vehicles (T3)		Useful for preventing or limiting the passage of vehicles (T3)
Installation type	Permanent	Permanent		Permanent

(continued)

Table 4.7 (continued)

Measure Class	Design of the physical elements of the BE [S1]		
Foundation	Rested on the ground/ pavement	Deep foundation	Extended and shallow
Anti-ram	Absent	Yes	Yes
Certificate/test	n.a	Vehicle 7 t; ≤ 80 km/h	Vehicle 7 t; ≤ 80 km/h
Source	[46]	[46]	[46]
Materials	Metals concrete, stone	Metals concrete, stone	Metals concrete, stone
Accessibility	Pedestrians, bicycles, wheelchairs	Pedestrians, bicycles, wheelchairs	Pedestrians, bicycles, wheelchairs
Integrability	Materials, shape	Materials, shape	Materials, shape
Interferences	Any	Urban surface network	Any
Efficacy (T2)	Not relevant	Not relevant	Not relevant
Efficacy (T3)	Low	High	High
OAs compatibility	Medium	Medium	Medium
Category	BOLLARDS [BO] 2/2		
SUB-CATEGORY	*[BO_4] INTEGRATED WITH FURNITURE*	*[BO_5] LUMINOUS*	*[BO_6] RETRACTILE*
Functioning type	Passive	Passive	Active
Description	Preventing or limiting the passage of vehicles (T3), combined with other urban furniture (e.g., bike rack)	Useful for preventing or limiting the passage of vehicles (T3)	Mobile for preventing or limiting the passage of vehicles (T3), when active
Installation type	Permanent	Permanent	Permanent
Foundation	Shallow foundation	Deep foundation	Deep foundation
Anti-ram	Yes	Yes	Yes
Certificate/test	n.a	ISO 179/1 eA = 70 kJ/ m^2	Vehicle 7 t; ≤ 80 km/h
Source	[44]	[44]	[46]
Materials	Metal	Metal, luminous device	Metals, concrete
Accessibility	Pedestrians, bicycles, wheelchairs	Pedestrians, bicycles, wheelchairs	Pedestrians, bicycles, wheelchairs
Integrability	Materials, functions	Materials, shape	Materials
Interferences	Any	Urban surface network	Urban surface network
Efficacy (T2)	Not relevant	Not relevant	Not relevant
Efficacy (T3)	Medium	High	High
OAs compatibility	Medium	Medium	Low
Measure Class	Design of the physical elements of the BE [S1]		
Category	INNOVATIVE SYSTEMS [IS] 1/2		

(continued)

Table 4.7 (continued)

Measure Class	Design of the physical elements of the BE [S1]		
SUB-CATEGORY	*[IS_1] ANTI-EXPLOSION FILM*	*[IS_2] BDP SYSTEM*	*[IS_3] BOMB JAMMER*
Functioning type	Passive	Passive	Active
Description	Useful device to make glass shatterproof. Used to reduce the possibility of glass shattering and therefore reduce the damage caused by the explosion. It is applied directly to existing glass	Device containing water to absorb the kinetic energy deriving from the impact of a vehicle, preventing the entire barrier from moving. Surrounding users may be flooded but not affected by the barrier (T2/T3)	Portable interference system useful for disabling the radio signal for the explosion of remotely controlled radio devices. Used to reduce the probability of a terrorist attack using radio-controlled explosives (T3)
Installation type	Permanent	Permanent/temporary	Not relevant
Foundation	Not relevant	Absent	Not relevant
Anti-ram	Yes	Yes	Not relevant
Certificate/test	ISO 616933 (EXV33C)	n.a	n.a
Source	Commercial product	BDP System patent	-
Materials	Plastic	Water, plastic	Electronic device
Accessibility	Not relevant	Pedestrians, bicycles, wheelchairs	Not relevant
Integrability	Only with glass	Shape	Not relevant
interferences	Any	Any	Radio devices
Efficacy (T2)	Not relevant	High	Not relevant
Efficacy (T3)	Medium	Medium	High
OAs compatibility	High	High	High
Category	INNOVATIVE SYSTEMS [IS] 2/2		
SUB-CATEGORY	*[IS_4] TURNTABLE BOLLARDS*	*[IS_5] METALLIC MESH*	
Functioning type	Active	Passive	
Description	Rotating system) useful for preventing the passage of vehicles (T3)	Device for preventing or limiting the passage of vehicles (T3), when positioned	
Installation type	Permanent	Temporary	

(continued)

Table 4.7 (continued)

Category	INNOVATIVE SYSTEMS [IS] 2/2			
Foundation	Shallow		Absent	
Anti-ram	Yes		Not	
Certificate/test	n.a		n.a	
Source	[46]		[44]	
Materials	Metal, concrete		Metal	
Accessibility	Pedestrians, bicycles, wheelchairs		Not relevant	
Integrability	Materials, shape		Not relevant	
interferences	Pavement		Any	
Efficacy (T2)	Not relevant		Not relevant	
Efficacy (T3)	Medium		Low	
OAs compatibility	Low		Low	
Measure class	BE layout [S2]		Safety and security management in the BE [S4]	
Category	SAFETY SIGNS [SS]		ALARM SYSTEMS [AS]	
SUB-CATEGORY	*[SS_1] LUMINOUS*	*[SS_2] STANDARD*	*[AS_1] MOBILE APP*	*[AS_2] PUBLIC ALARM SERVICE*
Functioning type	Active	Active	Active	Active
Description	Luminous road signs for indicating escape routes and safe points, even in low light conditions. Used to reduce the damage caused by a terrorist attack (T2/T3)	Road signs useful for indicating escape routes and safe points. Used to reduce the damage caused by a terrorist attack	System to transmit an emergency notification to mobile devices, road signs, radios, by authorities. Provides information and directions to follow in the event of a terrorist attack (T2/T3)	System that allows authorities to transmit a message (text message, email, road signs) to all devices in an emergency situation, providing information and directions to follow. Used to reduce the damage caused by a terrorist attack
Installation type	Permanent/ temporary	Permanent/ temporary	Not relevant	Not relevant
Foundation	Not relevant	Not relevant	Not relevant	Not relevant
Anti-ram	Not	Not	Not relevant	Not relevant

(continued)

Table 4.7   (continued)

Measure class	BE layout [S2]		Safety and security management in the BE [S4]	
Certificate/test	UNI EN ISO 7010:2012	UNI EN ISO 7010:2012	TS 102 900 V1.3.1	TS 102 900 V1.3.1
Source	UNI EN ISO 7010:2012	UNI EN ISO 7010:2013	TS 102 900 V1.3.1	TS 102 900 V1.3.2
Materials	Electronic device	Metal	Not relevance	Not relevant
Accessibility	Not relevant	Not relevant	Not relevant	Not relevant
Integrability	Not relevant	Not relevant	Not relevant	Not relevant
Interferences	Any	Any	Any	Any
Efficacy (T2)	Medium	Low	Medium	Medium
Efficacy (T3)	Low	Low	Medium	Medium
OAs compatibility	Medium	Medium	High	High
Measure class	Access control and surveillance in the BE [S3]			
Category	REMOTE CONTROL [RC]			
S\UB-CATEGORY	[RC_1] VIDEO SURVEILLANCE WITH AI	[RC_2] BIOMETRIC VIDEO SURVEILLANCE	[RC_3] VIDEO SURVEILLANCE TVCC	
Functioning type	Active	Active	Active	
Description	System that recognizes anomalies behaviours that signal the probability of an imminent crime. Employed to reduce the likelihood of occurrence of a terrorist attack (T2)	Biometric recognition system capable of identifying a person based on biological/ behavioural characteristics compared with data contained in a database (T2)	System designed to continuously record movements in the area of interest. images can be used to identify suspicious behaviour or reconstruct negative events (T2)	
Installation type	Permanent	Permanent	Permanent	
Foundation	Absent	Absent	Absent	
Anti-ram	Not	Not	Not	
Certificate/test	n.a	n.a	n.a	
Source	[44]	[47]	[48]	
Materials	Electronic device	Electronic device	Electronic device	
Accessibility	Not relevant	Not relevant	Not relevant	
Integrability	Shape and position	Shape and position	Shape and position	
interferences	Any	Any	Any	
Efficacy (T2)	High	High	Medium	
Efficacy (T3)	Not relevant	Not relevant	Not relevant	
OAs compatibility	High	High	High	

(continued)

Table 4.7 (continued)

Measure class	Access control and surveillance in the BE [S3]		
Category	DIRECT CONTROL [DC]		VIGILANCE [VG]
SUB-CATEGORY	[DC_1] IN TRANSIT METAL DETECTOR	[DC_2] MANUAL METAL DETECTOR	[VG] ARMED VIGILANCE
Functioning type	Active	Active	Active
Description	Device useful for detecting the presence of metal objects as users pass by. Used to reduce the probability of a terrorist attack using bladed weapons or firearms (T2)	Manual device useful for detecting the presence of metal objects. Used to reduce the probability of a terrorist attack using bladed weapons or firearms (T2)	Use of military personnel from the armed forces or public security forces with the function of controlling and supervising the built environment (T2/T3)
Installation type	Temporary	Temporary	Temporary/Permanent
Foundation	Not	Not	–
Anti-ram	Not	Not	–
Certificate/test	ISO 9001:2008	ISO 9001:2008	–
Source	ISO 9001:2008	ISO 9001:2009	National authorities
Materials	Electronic device	Electronic device	–
Accessibility	Not relevant	Not relevant	Controlled
Integrability	Any	Not relevant	Not relevant
Interferences	Any	Any	Not relevant
Efficacy (T2)	Medium	Medium	High
Efficacy (T3)	Not relevant	Not relevant	Medium
OAs compatibility	Low	High	Medium

References

1. Song Y, Liu B, Li L, Liu J (2022) Modelling and simulation of crowd evacuation in terrorist attacks. Kybernetes. https://doi.org/10.1108/K-02-2022-0260
2. Booth A, Chmutina K, Bosher L (2020) Protecting crowded places: challenges and drivers to implementing protective security measures in the built environment. Cities 107:102891. https://doi.org/10.1016/j.cities.2020.102891
3. Albores P, Shaw D (2008) Government preparedness: using simulation to prepare for a terrorist attack. Comput Oper Res 35:1924–1943. https://doi.org/10.1016/j.cor.2006.09.021
4. Bernardini G, Quagliarini E (2021) Terrorist acts and pedestrians' behaviours: first insights on European contexts for evacuation modelling. Saf Sci 143:105405. https://doi.org/10.1016/j.ssci.2021.105405
5. Quagliarini E, Fatiguso F, Lucesoli M et al (2021) Risk reduction strategies against terrorist acts in urban built environments: towards sustainable and human-centred challenges. Sustainability 13:901. https://doi.org/10.3390/su13020901
6. Quagliarini E, Bernardini G, Romano G, D'Orazio M (2023) Users' vulnerability and exposure in Public Open Spaces (squares): a novel way for accounting them in multi-risk scenarios. Cities 133:104160. https://doi.org/10.1016/j.cities.2022.104160

7. The European Commission (2022) Security by design: protection of public spaces from terrorist attacks

8. Han S, Song D, Xu L et al (2022) Behaviour in public open spaces: a systematic review of studies with quantitative research methods. Build Environ 223:109444. https://doi.org/10.1016/j.buildenv.2022.109444

9. Nemeškal J, Ouředníček M, Pospíšilová L (2020) Temporality of urban space: daily rhythms of a typical week day in the Prague metropolitan area. J Maps 16:30–39. https://doi.org/10.1080/17445647.2019.1709577

10. García-Palomares JC, Salas-Olmedo MH, Moya-Gómez B, et al (2018) City dynamics through Twitter: Relationships between land use and spatiotemporal demographics. Cities 72. https://doi.org/10.1016/j.cities.2017.09.007

11. Agliata R, Bortone A, Mollo L (2021) Indicator-based approach for the assessment of intrinsic physical vulnerability of the built environment to hydro-meteorological hazards: Review of indicators and example of parameters selection for a sample area. Int J Disaster Risk Reduct 58. https://doi.org/10.1016/j.ijdrr.2021.102199

12. Cains MG, Henshel D (2021) Parameterization framework and quantification approach for integrated risk and resilience assessments. Integr Environ Assess Manag 17:131–146. https://doi.org/10.1002/ieam.4331

13. Knowling MJ, White JT, Moore CR (2019) Role of model parameterization in risk-based decision support: an empirical exploration. Adv Water Resour 128:59–73. https://doi.org/10.1016/j.advwatres.2019.04.010

14. Zytoon MA (2020) A decision support model for prioritization of regulated safety inspections using integrated Delphi, AHP and double-hierarchical TOPSIS approach. IEEE Access 8:83444–83464. https://doi.org/10.1109/ACCESS.2020.2991179

15. Khodabocus S, Seyis S (2024) Multi-criteria decision-making model for risk management in modular construction projects. Int J Constr Manag 24:240–250. https://doi.org/10.1080/15623599.2023.2276649

16. Cantatore E, Quagliarini E, Fatiguso F (2024), Terrorism Risk Assessment for Historic Urban Open Areas, Heritage. 7:5319–5355. https://doi.org/10.3390/heritage7100251

17. Cantatore E, Esposito D, Sonnessa A (2023) Mapping the multi-vulnerabilities of outdoor places to enhance the resilience of historic urban districts: the case of the Apulian region exposed to slow and rapid-onset disasters. Sustainability 15:14248

18. Linstone HA, Turoff M (1975) The delphi method. Addison-Wesley Reading, MA

19. Villagràn De León JC (2006) Vulnerability: a conceptual and methodological review

20. IPCC (2012) Managing the risks of extreme events and disasters to advance climate change adaptation. Cambridge University Press, Cambridge

21. Koks EE, Jongman B, Husby TG, Botzen WJW (2015) Combining hazard, exposure and social vulnerability to provide lessons for flood risk management. Environ Sci Policy 47:42–52. https://doi.org/10.1016/j.envsci.2014.10.013

22. Quagliarini E, Bernardini G, D'Orazio M (2023) How could increasing temperature scenarios alter the risk of terrorist acts in different historical squares? a simulation-based approach in typological Italian squares. Heritage 6:5151–5188

23. Beňová P, Hošková-Mayerová Š, Navrátil J (2019) Terrorist attacks on selected soft targets. J Secur Sustain Issues 8:453–471. https://doi.org/10.9770/jssi.2019.8.3(13)

24. Ministry of Interior (Italy) (2015) DM 03/08/2015: Fire safety criteria (Approvazione di norme tecniche di prevenzione incendi, ai sensi dell'articolo 15 del decreto legislativo 8 marzo 2006, n. 139.)

25. Bernardini G, Ferreira TM, Julià PB et al (2024) Assessing the spatiotemporal impact of users' exposure and vulnerability to flood risk in urban built environments. Sustain Cities Soc 100:105043. https://doi.org/10.1016/j.scs.2023.105043

26. Bloomberg M, Burden A (2006) New York City pedestrian level of service study-phase 1. NY, USA, New York

27. Bosina E, Weidmann U (2017) Estimating pedestrian speed using aggregated literature data. Phys A Stat Mech its Appl 468:1–29. https://doi.org/10.1016/j.physa.2016.09.044

28. Haghani M, Kuligowski E, Rajabifard A, Lentini P (2022) Fifty years of scholarly research on terrorism: Intellectual progression, structural composition, trends and knowledge gaps of the field. Int J Disaster Risk Reduct 68:102714. https://doi.org/10.1016/j.ijdrr.2021.102714
29. van der Wal CN, Formolo D, Robinson MA, et al (2017) Simulating crowd evacuation with socio-cultural, cognitive, and emotional elements. Lect Notes Comput Sci (including Subser Lect Notes Artif Intell Lect Notes Bioinformatics) 10480 LNCS:139–177. https://doi.org/10.1007/978-3-319-70647-4_11
30. De Lotto R, Pietra C, Venco EM (2019) Risk analysis: a focus on urban exposure estimation. Computational science and its applications—ICCSA 2019. Springer, Cham, pp 407–423
31. Hostetter H, Naser MZ (2022) Characterizing disability in fire: a progressive review. J Build Eng 53:104573. https://doi.org/10.1016/j.jobe.2022.104573
32. Ronchi E, Kuligowski ED, Reneke PA, et al (2013) The process of verification and validation of building fire evacuation models. NIST Tech Note 1822:
33. Quagliarini E, Bernardini G, Romano G, D'Orazio M (2022) Simplified flood evacuation simulation in outdoor built environments. Preliminary comparison between setup-based generic software and custom simulator. Sustain Cities Soc 81:103848. https://doi.org/10.1016/j.scs.2022.103848
34. Lu P, Wen F, Li Y, Chen D (2021) Multi-agent modeling of crowd dynamics under mass shooting cases. Chaos, Solitons Fractals 153:111513. https://doi.org/10.1016/j.chaos.2021.111513
35. Gravetter FJ, Wallnan LB (2013) Statistics for the behavioural sciences, 9th edn. Wadsworth, Cengage Learning
36. Liu Q (2018) A social force model for the crowd evacuation in a terrorist attack. Phys A Stat Mech its Appl. https://doi.org/10.1016/j.physa.2018.02.136
37. Ronchi E, Uriz FN, Criel X, Reilly P (2015) Modelling large-scale evacuation of music festivals. Case Stud Fire Saf 5:11–19. https://doi.org/10.1016/j.csfs.2015.12.002
38. Bernardini G, Ferreira TM (2022) Emergency and evacuation management strategies in earthquakes: towards holistic and user-centered methodologies for their design and evaluation. In: Ferreira TM, Rodrigues H (eds) Seismic vulnerability assessment of civil engineering structures at multiple scales. Woodhead Publishing - Elsevier, pp 275–321
39. Guo K, Zhang L, Wu M (2023) Simulation-based multi-objective optimization towards proactive evacuation planning at metro stations. Eng Appl Artif Intell 120:105858. https://doi.org/10.1016/j.engappai.2023.105858
40. Syed Abdul Rahman SAF, Abdul Maulud KN, Pradhan B et al (2021) Impact of evacuation design parameter on users' evacuation time using a multi-agent simulation. Ain Shams Eng J 12:2355–2369. https://doi.org/10.1016/j.asej.2020.12.001
41. Bernardini G (2017) Fire safety of historical buildings. Traditional Versus Innovative "Behavioural Design" Solutions by Using Wayfinding Systems, 1st ed. Springer International Publishing
42. Banerjee A, Maurya AK, Lämmel G (2018) Pedestrian flow characteristics and level of service on dissimilar facilities: a critical review. Collect Dyn 3:A17. https://doi.org/10.17815/CD.2018.17
43. Coaffee J, Bosher L (2008) Integrating counter-terrorist resilience into sustainability. Proc Inst Civ Eng Des Plan 161:75–83
44. GCDN Commissioned Research (2018) Beyond concrete barriers innovation in urban furniture and security in public space
45. Cantatore E, Quagliarini E, Fatiguso F (2022) European cities prone to terrorist threats: phenomenological analysis of historical events towards risk matrices and an early parameterization of urban built environment outdoor areas. Sustainability 14. https://doi.org/10.3390/su141912301
46. Federal Emergency Management Agency (2007) FEMA 430: site and urban design for security: guidance against potential terrorist attacks
47. Lyon D (2008) Biometrics, identification and surveillance. Bioethics 22:499–508
48. (NaCTSO) NCTSO, Kingdom U, Infrastructure C for the P of N, Kingdom U (2012) Protecting Crowded Places: Design and Technical Issues

Chapter 5
A Case Study Application: Vittorio Veneto Square in Matera, Italy

Abstract The chapter applies the theories and methods for terrorist risk assessment and behavioural analysis presented in this book to the peculiar case study of Vittorio Veneto Square in Matera, a city in the Basilicata region located in the south of Italy. This outdoor Open Area (OA) is representative in view of the presence of several special buildings, defining a high potential level of attractiveness for terrorist acts. Moreover, the square is characterized by a high level of tourist attraction for the strategic position near the "Sassi", the UNESCO site of Matera, and this condition increases the relevance as a soft target because of significant users' exposure. Scenarios for risk assessment are first created, and then behavioural-based assessment is performed thanks to a validated simulation model, considering the current conditions of the square. Scenarios referring to the evacuation of the square (without interactions between the crowd and the perpetrators) are compared with those related to an armed assault with cold weapons, using behavioural-based key performance indicators. Then, selected mitigation strategies based on emergency planning, and thus compatible with the cultural and historical relevance of the place, have been considered and tested according to the same approach. Applying the proposed approach is expected to support decision-makers and, mainly, local administrations while evaluating the OAs resilience towards terrorist acts, thus boosting the risk assessment and mitigation planning.

Keywords Behavioural design · Case study · Simulation · Risk assessment · Risk mitigation · Terrorist acts · Outdoor open areas

5.1 The Case Study: Vittorio Veneto Square in Matera, Italy

The selected case study to apply behavioural design methods for terrorist risk assessment and mitigation concerns the outdoor Open Area (OA) of Vittorio Veneto Square in Matera, Italy. Vittorio Veneto Square is a public square located within the actual perimetration of the historic district of Matera, a city in the Basilicata region in the

G. Bernardini et al., *Terrorist Risk in Urban Outdoor Built Environment*,
SpringerBriefs in Architectural Design and Technology,
https://doi.org/10.1007/978-981-97-6965-0_5

south of Italy. Its development is related to the urban expansion from the thirteenth to the seventeenth centuries, when the actual buildings were built, while its physical transformations and maintenance rely on its uses. It represents the nodal point for tourist access to the Sassi, the complex underground system of ancient dwellings of Matera listed as a UNESCO site [1].

Vittorio Veneto Square is configured as a flat terrace overlooking the Sassi, albeit, in the transformation process, direct exposure to the Sassi has been restricted to some points due to the construction of buildings along its perimeter (Fig. 5.1). Two main un-built elements allow to connect the square to Sassi: the balcony, for the landscape view of Sassi, created within a Loggia ("Belvedere Luigi Guerricchio") in the central part of the western-built profile, and stairs to physically connect the square to Sassi. Another physical discontinuity of the floor is represented by the presence of the "Palombaro", ancient water cisterns dug in 1882 under the square and connected with the other cisterns in the Sassi to serve the city. They fell into disuse with the activation of the urban aqueduct in 1991 but today they are uncovered to be observed from the square as engineered masterpieces, protecting their perimeter with fences [1]. Similarly, a second archaeological dug interrupts the pavement of Piazza Vittorio Veneto, allowing tourists to observe lower rooms while fences limited the perimeter. The Loggia and the "Palombaro" are both placed, in the northern part of the OA.

The built profile reflects the historical evolution of the square, presenting private palaces, with 2–3 floors, and public buildings. Most of the ground floors of private dwellings host commercial activities (bars, pubs, restaurants, shops), while public buildings urban services. Specifically for these, three main public uses can be identified:

- A bank in the southern part of the square.
- A public library and a cinema-theatre within the Palazzo dell'Annunziata, in the northern part.
- The prefecture, which is hosted by the ex-convent of the San Domenico Church, in the northern part.

Moreover, the square presents also a religious fabric, the San Domenico Church.

The case study OA is characterized by a complex morphology, merging a main trapezoidal shape in the northern part of the square to a rectangular one in the southern. Concerning the square connection with the surrounding urban built environment, six access streets can be recognized, mainly pedestrian (Fig. 5.2), limiting the vehicular traffic within the same pedestrian square. However, three possible vehicular access streets can be identified: Via Roma, Via del Corso, Via Luigi Lavista (Fig. 5.2). Here, the geometric dimension of entrances and the urban vehicular traffic regulations allow to move close to the square, while fixed bollards limit the vehicular accessibility. In view of the above, it is worth noting that the assessed OA is also characterized by multiple uses, combining the public and touristic services and attractions that make Vittorio Veneto Square a sensitive soft target. In fact, the symbolism of the place determined by the presence of the UNESCO site and the presence of high touristic (cultural) flux determine a high potential level of proneness. This can be summarized

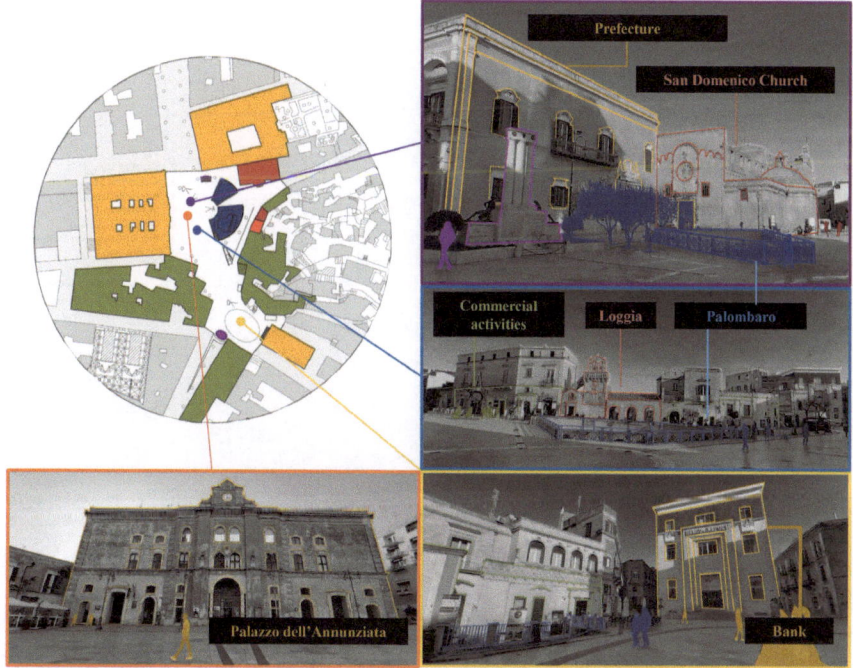

Fig. 5.1 Relevant (symbolic, strategic and commercial) buildings and areas in Piazza Vittorio Veneto (photos of the authors)

as the local relevance of the place by local users which usually serve public places and buildings for daily activities. Moreover, the tourist inflow has increased during the last years due to the international relevance of Matera as the "2019 European Capital of Culture" which enhances the relevance of the place for tourist attractiveness and symbolism [2].

Methods and tools provided in Chap. 4 are then used to create scenarios and evaluate risk due to terrorist acts in the case study, using the same phase order. In particular, the risk assessment methods devoted to provide possible attack points are applied in Sect. 5.2, while the time-dependent assessment of the users' exposure and vulnerability is defined in Sect. 5.3. According to the outcomes of these methodological steps, simulations are finally performed to investigate risk levels in different pre- and post-retrofit scenarios, as shown in Sect. 5.4, using behavioural and simulation-based key performance indicators (KPIs) to compare evacuation issues under the given boundary conditions. To this end, selected mitigation strategies implemented in post-retrofit scenarios refer to emergency management and planning, thus ensuring the best compatibility with the heritage features of Vittorio Veneto Square, without altering the layout or the identity of the places.

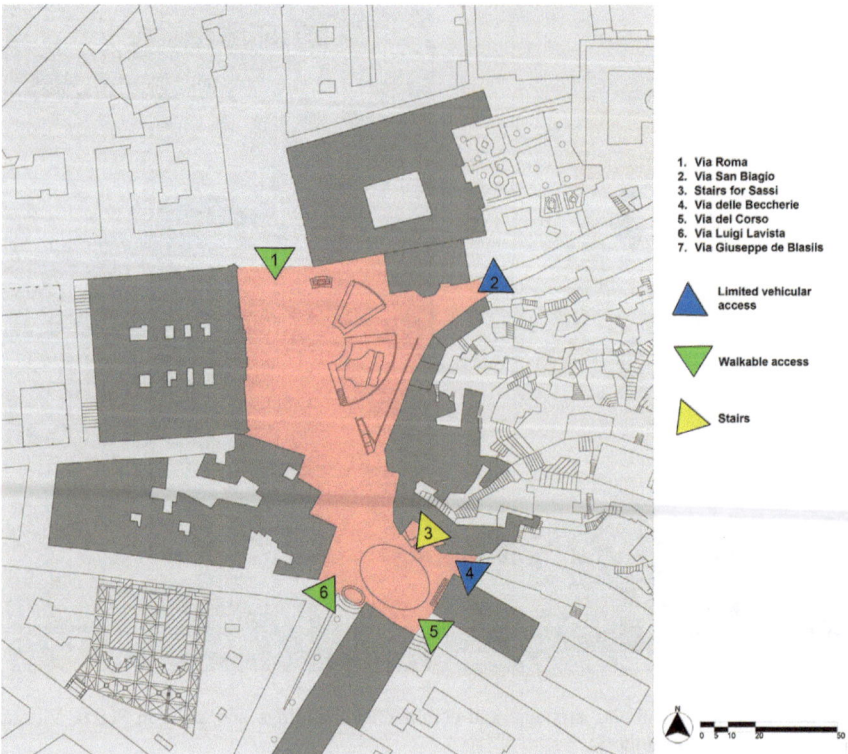

Fig. 5.2 Morpho-typological shape of Vittorio Veneto Square and characterization of accesses (Base map from CTR of Basilicata)

5.2   Risk Assessment of OAs to Provide Possible Attack Points: Pre-Retrofit Scenarios

Starting from the terrorism risk assessment formulation in OAs defined in Chap. 4, Sect. 4.2 and based on the approach of previous works of the authors [3], all the required data are gathered and qualified. Considering the building intended uses presented in the previous section, their spaces of relevance (SoRs) are calculated and modelled in the plan (Fig. 5.3).

The buildings and attractive places considered in the analysis include special and public buildings, tourist attractive places such as "Palombaro" and Loggia on the Sassi, public commercial activities such as bars, their covered terraces, and stores. Specifically, for buildings, the SoRs have been calculated considering their commercial extension and the number of floors occupied for the use [m^2], which is also expressed by GS$_i$ (compare with Sect. 5.3); while for all the gathering elements and parts of the squares, such as covered or uncovered bar terraces, their SoR extension

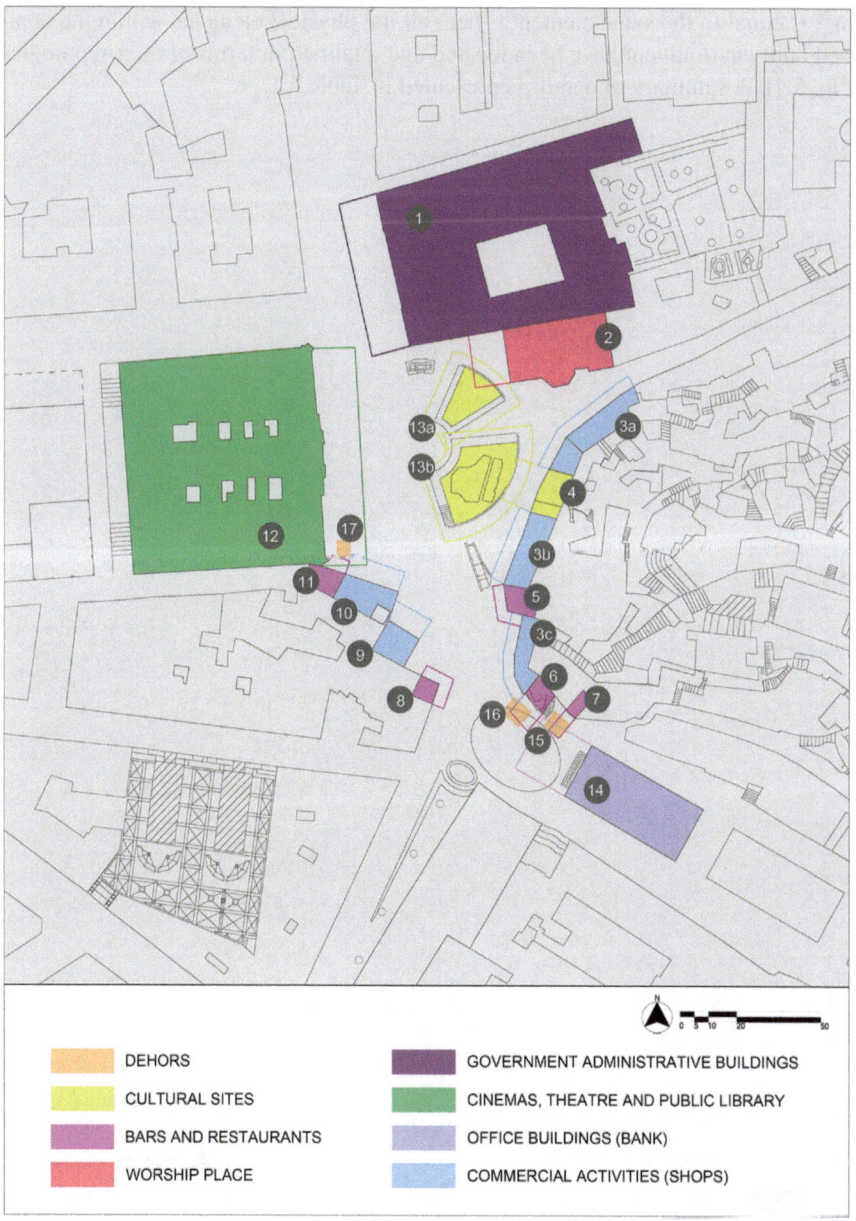

Fig. 5.3 Identification of outdoor and indoor intended uses in Vittorio Veneto Square and perimetration of associated Spaces of Relevance (SoRs), according to Chap. 4 methods (Base map from CTR of Basilicata)

[m^2] is equal to the same element. Then, all the physical elements within the analysed built environment have been located and qualified in terms of their typologies (Fig. 5.4). A summary of details is presented in Table 5.1.

COVERED BAR TERRACES		BOLLARD	
STAIR		TREE	
MONUMENT		LAMPPOST	
FLOWERPOT		FENCES	

Fig. 5.4 Position and classification of physical objects in Vittorio Veneto Square (Base map from CTR of Basilicata)

Table 5.1 Extension of public and strategic buildings and associated SoRs, classified coherently with the Classes of Built Environment defined in Chap. 3 (F, F_B, and F_D) and extension of potential gathering areas (dehors)

ID	Building/area of interest	GS_i [m^2]	Extension of SoR [m^2]	Type of CBE
1	Prefecture—ex Convent	2700	270	F_D
2	San Domenico Church	350	245	F_B
3a	Shops	70	28	F_B
3b	Shops	50	20	F_B
3c	Shops	50	20	F_B
4	Loggia Luigi Guerricchio	60	72	F_B
5	Restaurant	50	60	F_B
6	Restaurant	60	72	F_B
7	Bar	35	42	F_B
8	Bar	30	36	F_B
9	Shops	110	77	F_B
10	Shops	70	49	F_B
11	Bar	20	24	F_B
12	Library and cinema-theatre	1330	742	F_B
13a	Archaeological dig	120	84	F_B
13b	Palombaro	210	147	F_B
14	Bank	580	232	F_B
15	Dehor	27	–	F_B
16	Dehor	50	–	F_B
17	Dehor	25	–	F_B
18	Square (whole pedestrian area)	5000	–	F

The results of risk calculations for each area and SoR are processed considering the attack types recognized for OAs, T2—armed assault and T3—bombing attack with a vehicle, summarized in Tables 5.2 and 5.3 and outlined in Figs. 5.5 and 5.6.

As a critical analysis of results, two main aspects can be discussed:

- Coherently with Chaps. 2 and 4, Sect. 4.2, the analysed OA has a major proneness to T2. The presence of physical objects and vehicular traffic regulations allow the reduction of the proneness to T3 performed with a vehicle. This is also enhanced for Matera for its intrinsic closeness, presenting a reduced number and geometric extension of accesses. This is clearly shown in Figs. 5.5 and 5.6 where the distribution of SoRs featured by a medium risk level is distributed along the northern part of the squares (Prefecture, Teather) where the largest access is located. Even if this access is featured by the presence of surface bollards, a possible scenario can provide external attacks to the vehicles, towards such peripherical areas. When the focus is on the southern part, the relevance of the risk levels for public uses

Table 5.2 Risk determinant values resulting from the application of the algorithm for each SoR and area determined in Vittorio Veneto Square, for the attack Type T2

ID	Building/area of interest	H	V	E	Level of risk
1	Prefecture – ex Convent	3	4	2	Medium
2	San Domenico Church	3	4	3	Medium
3a	Shops	2	3	2	Negligible
3b	Shops	2	3	2	Negligible
3c	Shops	2	3	2	Negligible
4	Loggia Luigi Guerricchio	3	3	4	Medium
5	Restaurant	2	3	3	Medium
6	Restaurant	2	4	4	High
7	Bar	3	4	4	High
8	Bar	3	3	4	Medium
9	Shops	2	3	2	Negligible
10	Shops	2	3	2	Negligible
11	Bar	3	4	4	High
12	Library and cinema-theatre	3	3	3	Medium
13a	Archaeological dig	3	3	3	Medium
13b	Palombaro	3	3	3	Medium
14	Bank	3	2	2	Low
15	Dehor	3	4	3	Medium
16	Dehor	3	4	3	Medium
17	Dehor	3	4	3	Medium
18	Square (whole pedestrian area)	4	4	5	High

(bars, restaurants, and shops) is reduced due to the presence of physical objects (trees and lampposts) that enhance the local protection of users.

- Instead, the T2 attack type provides multiple scenarios. Despite the physical accessibility by perpetrators in all the places of the square, the proneness of SoRs and area is mainly determined by the potential crowding levels, while the global riskiness is related to the intrinsic vulnerability determined by the presence of obstacles. That is clear in the reading of results outlined in Fig. 5.5 where medium and high-risk SoRs overlap nearest the touristic attractive places (Palombaro and Loggia) or where densely crowded uses merged with extended obstacles (dehors).

This analytic reading of the phenomenon through qualitative and quantitative data allows to interpret the OA and to determine two possible attack points for the T2 type while neglecting the T3 one, as schematized in Fig. 5.7.

Specifically, the determined attack points describe two possible significant scenarios coherently with the terroristic strategies and efficacy of violent acts:

Table 5.3   Risk determinant values resulting from the application of the algorithm for each SoR and area determined in Vittorio Veneto Square, for the attack Type T3

ID	Building/area of interest	H	V	E	Level of risk
1	Prefecture—ex Convent	4	3	4	Medium
2	San Domenico Church	3	2	2	Low
3a	Shops	2	2	2	Negligible
3b	Shops	2	2	2	Negligible
3c	Shops	2	2	2	Negligible
4	Loggia Luigi Guerricchio	2	2	3	Low
5	Restaurant	2	2	3	Low
6	Restaurant	2	2	3	Low
7	Bar	2	2	3	Low
8	Bar	3	3	4	Medium
9	Shops	3	3	2	Low
10	Shops	3	3	2	Low
11	Bar	2	3	2	Negligible
12	Library and cinema-theatre	3	3	3	Medium
13a	Archaeological dig	3	2	2	Low
13b	Palombaro	3	2	2	Low
14	Bank	3	3	2	Medium
15	Dehor	2	4	2	Negligible
16	Dehor	2	4	2	Negligible
17	Dehor	2	3	2	Negligible
18	Square (whole pedestrian area)	3	3	4	Medium

- Scenario 1 (AS1) describes a possible attack which aims at maximizing both the number of people involved and the media publicity, involving two major touristic and cultural places within the square (Palombaro and Loggia).
- Scenario 2 (AS2) illustrates one of the most recurrent attack cases, where the aim is the maximization of the effect striking some of the most crowded areas also featured by very low protective elements (bar, pub).

5.3   Mitigation Strategies Identification: Post-Retrofit Scenarios

As discussed in Chap. 4, the identification of strategies to prevent and mitigate the effects of a terroristic attack is strictly linked to the attack type and modus operandi. The identification of possible scenarios of violent acts to be carried out through cold arms for the case study of Vittorio Veneto Square requires to be merged with the

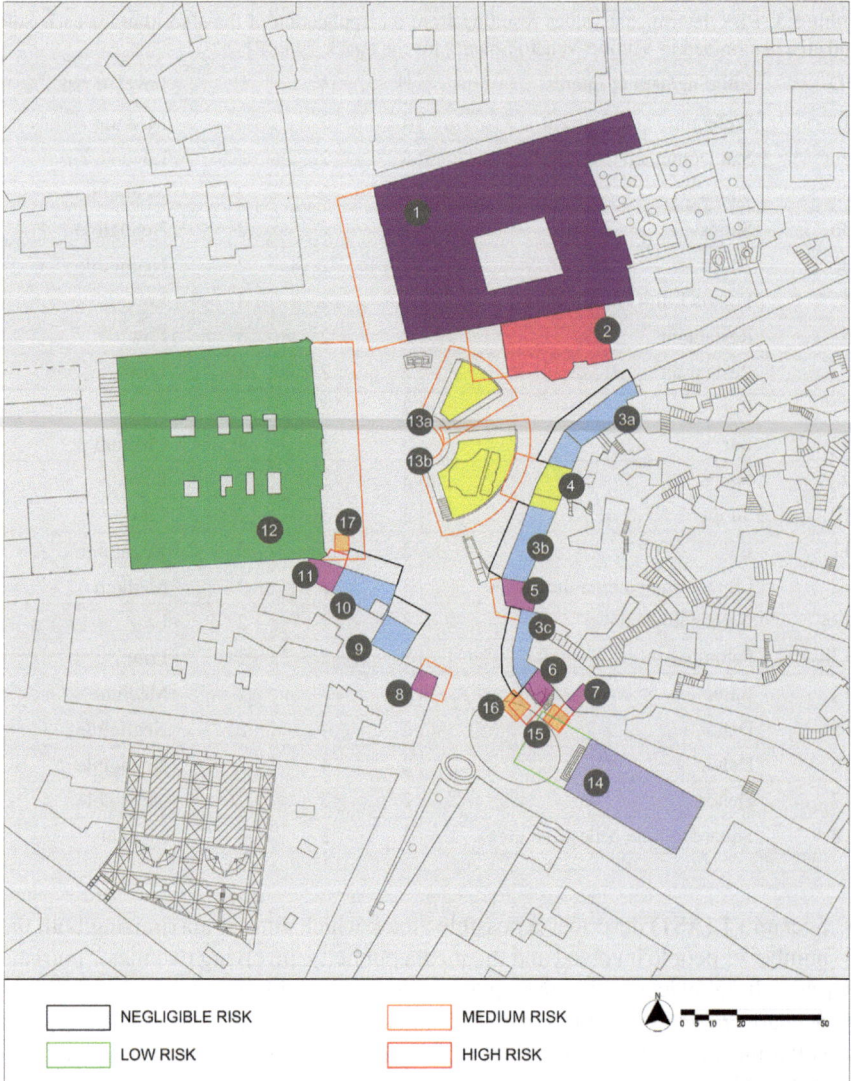

Fig. 5.5 Risk levels for T2 attack type applied to SoR and area extension provided in Table 5.1 and translated in Red, orange, green, and black lines to discuss the high, medium, low, and negligible levels of risks, calculated and reported in Table 5.2. (Base map from CTR of Basilicata)

possible effective strategies and their efficacy. However, due to its inherent features, the strategies for the T2 attack type have a prevalent tactical dimension. In fact, even if literature, guidelines, and previous experiences have highlighted the relevance of physical obstacles in determining possible temporary secure areas during the violent

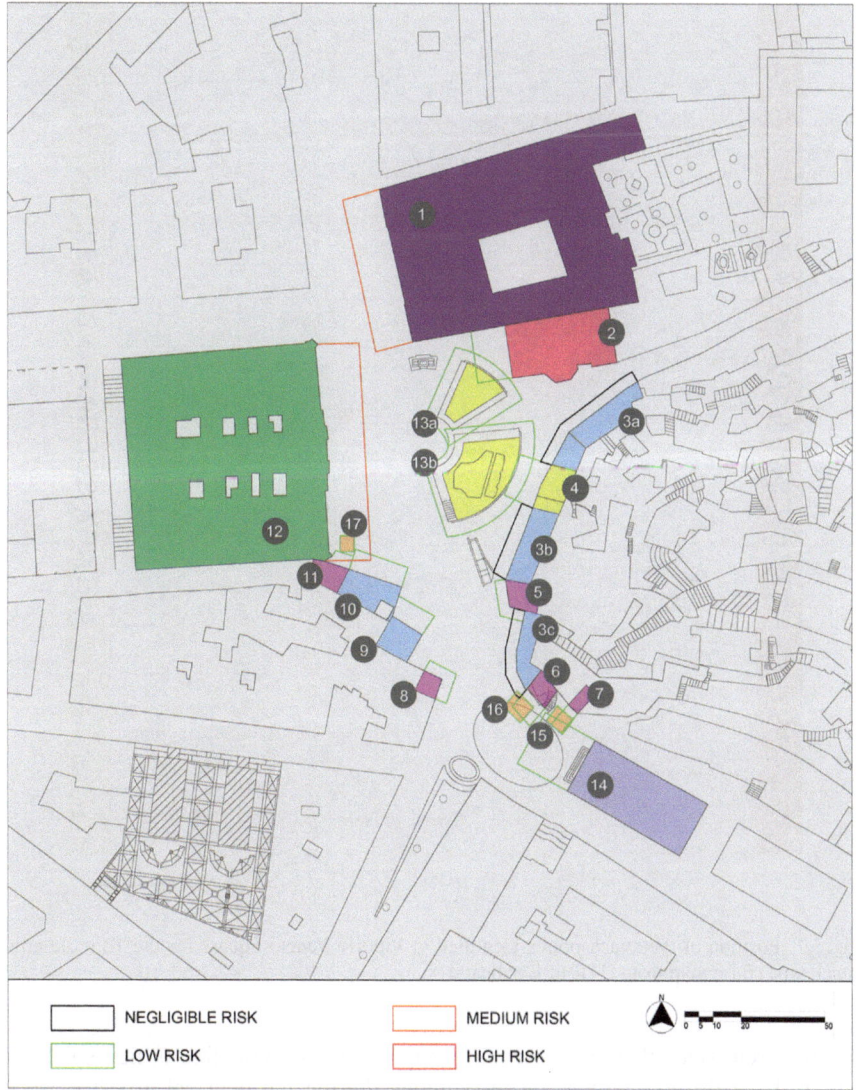

Fig. 5.6 Risk levels for T3 attack type applied to SoR and area extension provided in Table 5.1 and translated in red, orange, green, and black lines to discuss the high, medium, low, and negligible levels of risks, calculated and reported in Table 5.3 (Base map from CTR of Basilicata)

act, such effectiveness requires to be combined with coherent actions of education of users.

Starting from the Risk Mitigation and Reduction Strategies (RMRSs) classification (Chap. 2, Sect. 2.3), strategies related to security personnel deployment, emergency management, and wayfinding in emergency scenarios have been selected in

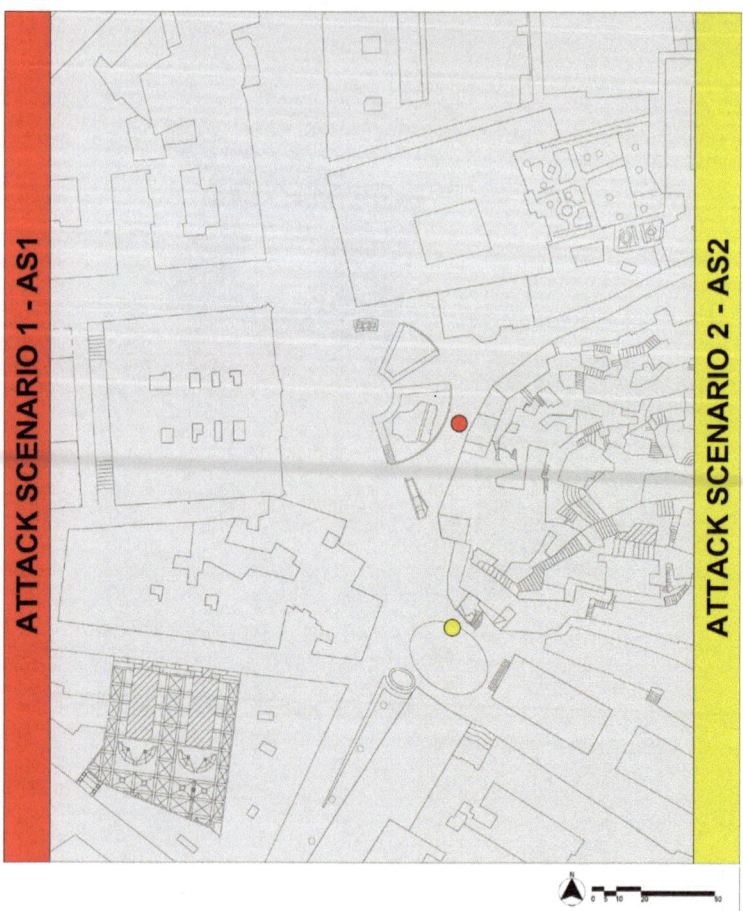

Fig. 5.7 Position of the attack points identified in Vittorio Veneto Square for the T2 (cold arm) attack type (Base map from CTR of Basilicata)

this research, since they can essentially support users and define design solutions within the built environment layout without altering the physical aspect of such places. On the other side, these strategies can be also consistent with the supposed "modus operandi" in the T2 attack, which does not imply damage to the buildings, thus limiting the need for interventions on physical elements, and which contrarily implies a dynamic perpetrator behaviour according to the prey-predator modelling criteria (see Chap. 3, Sect. 3.4). Considering the peculiarities of the assessed case study, moreover, the nearby presence of the local prefecture may support this kind of action. Near to that, the study of possible evacuation scenarios involving different conditions of evacuation with respect to the accesses and the LEA's position can support the study and the expectation of emergency management.

Due to that, post-retrofit scenarios are outlined merging the configuration of the real investigated OA and the outlined attack points, preparing the simulation scenarios for the qualification of single strategies when combined (Figs. 5.8, 5.9, and 5.10). Specifically, for Scenario 1 (AS1), the strategy involves LEA in the northern part of Vittorio Veneto Square (AS1-ST1), taking advantage of the Prefecture (Fig. 5.8); in Scenario 2 (AS1) the first strategy considers closing minor exits (in the southern part) to the advantage of more wide accesses in the northern part (AS2-ST1) (Fig. 5.9), while the second adds the LEA in the nearest of the Prefecture as a temporary secure area to be reached (AS2-ST2) (Fig. 5.10).

5.4   Time-Dependent Assessment of User-Related Factors

The time-dependent assessment of user-related factors has been performed according to Chap. 4, Sect. 4.3 methods, using remote analysis and quick (standard) input data to focus on the rapid application of the methodology and the related capability demonstration.

The main areas connected to outdoor and indoor intended uses which can generate overcrowding are shown in Fig. 5.3, and selected according to the classification of Chap. 4, Table 4.3. The main exits/access streets to the OA considered in the evacuation process and the main obstacles outdoors (that are greeneries, fountains, outdoor walls, and street furniture) are shown in Fig. 5.4.

Intended uses are then associated with their related main features in terms of surface, users' typologies (only outdoor users—OO, prevalent outdoor users—PO, non-residents—NR), and users' exposure over time. From a methodological perspective, the intended uses are associated with the OA via Google Maps/Street, the available gross surface GS_i [m^2] is calculated by Calcmaps, and standard occupant loads and online timetables are assumed to pursue the rapid applicability of the proposed methodology, thus avoiding time-consuming on-site survey. For these reasons, uncertainties due to specific conditions of the square and to the relationship of the OA with the whole urban fabric (e.g., in terms of visitors' flows) can exist, but the whole capabilities of the methodology are not affected by such simplified assumptions. The users' vulnerabilities assessment by users' age and gender are also performed according to a quick approach, thanks to the statistics of the National ISTAT annual reports,[1] assuming a homogeneous municipality-related distribution of data for the sake of simplicity.

In view of the, Table 5.4 traces the summary of the main features of the intended uses open to the public, thus excluding residential areas as in the rationale of the methodology in view of the attack attraction towards soft targets [4, 5]. Particular attention is paid to NR associated with special buildings having a symbolic value or that are widely interested in visitors' presence over the day (marked with * in Table 5.4), since their position can be associated with the immediate outdoor areas of

---

[1] https://demo.istat.it/app/?i=POS&a=2023&l=en; last access: 20/02/2024.

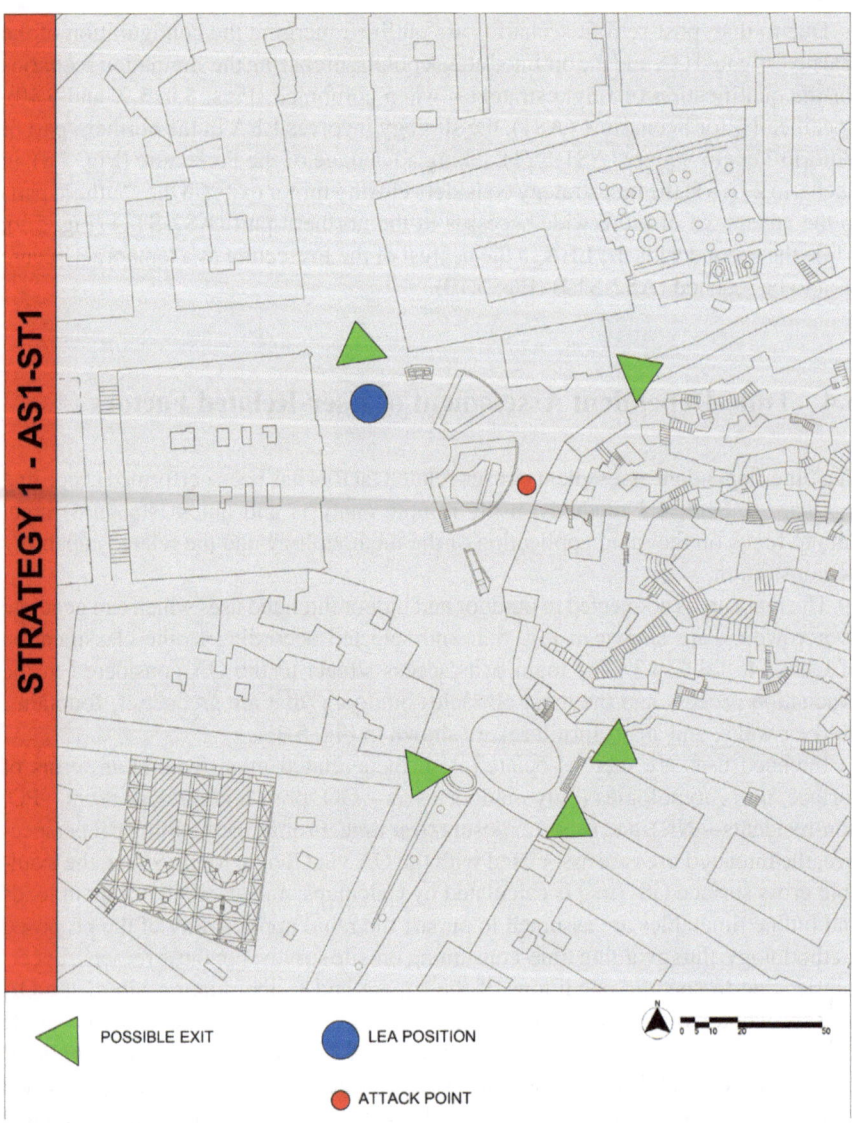

Fig. 5.8 Details of strategy 1 (ST1) applied in Scenario 1 (AS1) of the attack type T2, highlighting the position of LEA and possible exits (Base map from CTR of Basilicata)

the buildings, thus maximizing exposure [6] (compare Chap. 3, Sect. 3.4 on the simulation model). Therefore, the total number of users per intended use $NU_{t,i}$ [persons] is also shown in Table 5.4. Figure 5.11 shows the trends of the main user-related KPIs defined in Chap. 4, Table 4.4, focusing on occupant density (Fig. 5.11a), normalized KPIs on overall exposure and exposure referring to the outdoors (Fig. 5.11b), percentages of users by use behaviours (Fig. 5.11c) and age (Fig. 5.11d). In this sense,

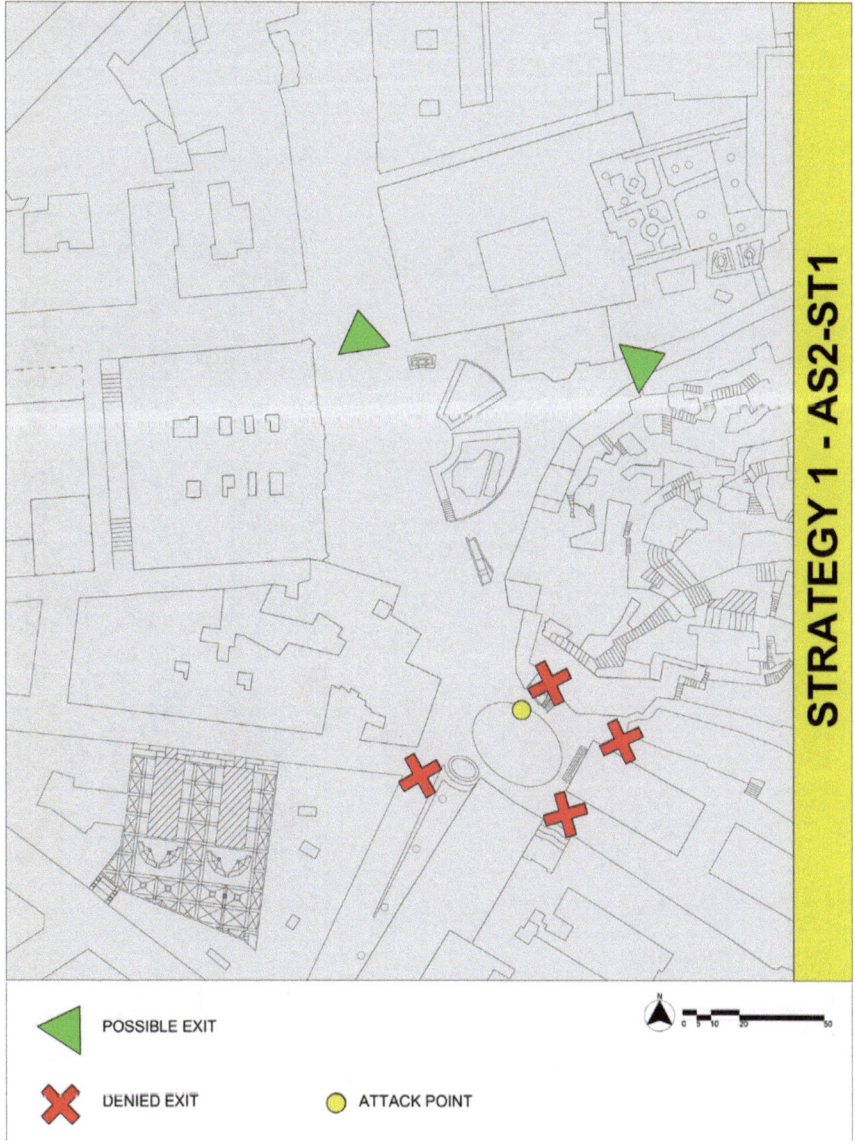

Fig. 5.9 Details of strategy 1 (ST1) applied in Scenario 2 (AS2) of the attack type T2, highlighting possible exits (Base map from CTR of Basilicata)

Fig. 5.10 Details of strategy 2 (ST2) applied in Scenario 2 (AS2) of the attack type T2, highlighting the position of LEA and possible exits (Base map from CTR of Basilicata)

according to Table 5.1, working days seem to be related to higher users' exposure levels due to the higher crowding. Thus, the data shown in Fig. 5.11 refers to typical working days.

Considering the overall users' outdoor density in outdoor at a given time t UOd_t [persons/m^2] (Fig. 5.11a) case study, OA seems to be characterized by the highest exposure values during the late morning and the afternoon. The maximum exposed users' density (about 0.37 persons/m^2) is reached at about $t = 18$, but values are lower than those of the whole users' density, also comprising users placed indoors and in possible protected areas (about 0.73 persons/m^2) at the same given daytime. As expected, this hour of the day is affected by the highest users' normalized number NUn_t [-], as shown by Fig. 5.11b. Nevertheless, the effects of users' exposure in the outdoors seem to be more evident when the number of NR placed indoors is minimized, i.e. during the nighttime and in the late evening, as remarked by the KPI relating to the impact of an event in the OA on the whole population at a given time t, that is IE$_t$ [-] (Fig. 5.11b). Therefore, the analysis of Fig. 5.11a, b should be jointly performed since the two panels and the related KPIs show different aspects of the exposure assessment in the OA. The most crowded scenario refers to $t = 18$, indeed, when NU$_t$ is about 3900 persons (that is 0.73 persons/m^2 $\times$ 5390 m^2 of outdoor surface including pedestrian areas and other sites, according to Table 5.4) and the effective number of users placed outdoors, and thus exposed to the attack (summing OO, PO, NR considered as special buildings visitors as in Table 5.1), is about 1940 persons. On the contrary, at $t = 23$, IE$_t = 0.87$ but $UOd_t = 0.32$ persons/m^2, and thus, the overall number of effective users exposed to the attack is significantly lower.

In addition, Fig. 5.11c then traces the impact of OO, PO, and NR (also distinguishing between visitors of special buildings), so as to point out possible dynamics in the use behaviours. Finally, Fig. 5.11d outlines the percentage of users by age, that is toddlers T (0–4 years), parents-assisted children PA (5–14 years), young autonomous YA users (15–19 years), adult users AU (20–69 years) and elderly users EU (70 + years), showing that they are almost constant over the daytime. This result is essentially affected by the quick assessment approach in users' vulnerability by age relying on homogeneous statistical data.

In view of the above, it is hence possible to conclude that:

- The most critical scenario in terms of users' exposure and vulnerability is related to $t = 18$, essentially in view of the higher density of users and thus the number of possible involved individuals affected by the attack. This scenario will be used for generating simulation inputs;
- Nevertheless, evening time scenarios are still critical since most of the users in the OA are placed outdoors, although the overall density is lower than the one in the afternoon and in the late morning.
- The contribution due to visitors of special buildings is mainly significant during the late afternoon and evening times (Fig. 5.11c), and thus considering them as placed outdoors could support a conservative approach to risk assessment.
- The users' vulnerability seems to have a limited influence on the whole assessment process, but this outcome can be checked by in situ surveys. Surveys can also

Table 5.4 Intended uses of public areas for the case study according to Fig. 5.4 identification of spaces in the case study OA, by characterizing the overall available gross surface GS_i, the behaviour of the hosted users, the quick occupant loads OL_i according to Chap. 4, Table 4.4, the total number of users in the OA NU_t, opening times and notes

Intended use	GS_i [m²]	Use behaviours	OL_i [persons/ m²]	$NU_{t,i}$ [persons]	Timetable (open to public) and notes: working W and holiday H reference for timetable
Pedestrian areas	5000	OO	0.1	500	-
Dehors	102	PO	0.4	41	10AM-11PM: W & H
Other sites (potential outdoor mass gatherings)	390	OO	0.4	156	crowding distributed to the whole pedestrian area as visitors (passersby): W & H
Bars and restaurants	195	NR	0.7	137	10AM-11PM: W & H
Worship place	350	NR*	0.7	245	8–12 AM and 3-7PM: W & H
Government administrative buildings	2700	NR	0.1	270	9AM-5PM, mainly offices closed to public: W
Cinemas, theatre	700	NR*	1.2	840	6-10PM: W & H
Public library	630	NR*	0.2	126	9AM-5PM: W
Office building (bank)	580	NR	0.4	232	8AM-1PM and 2-4PM: W
Shops, other commercial buildings	350		0.4	140	10AM-10PM: W & H
Unwalkable areas/ monuments and obstacles	423	-	0	0	-

*: NR relates to visitors of special buildings

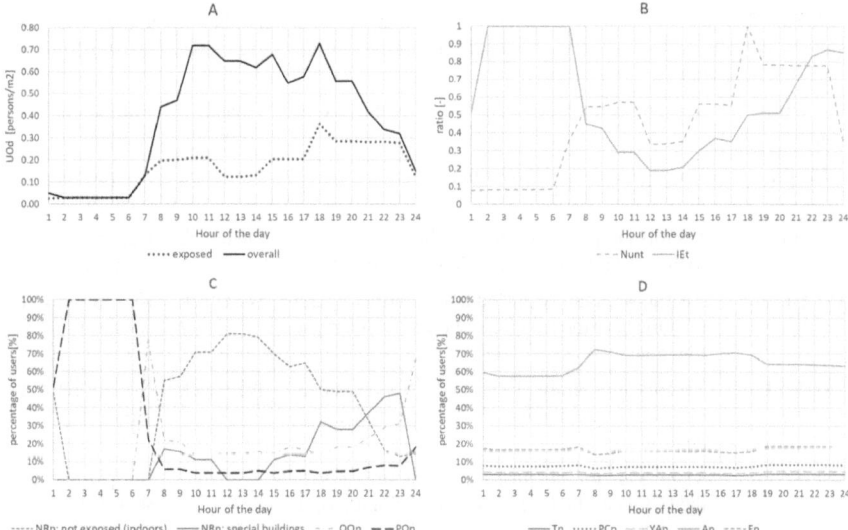

Fig. 5.11 KPIs trends over daytime considering users' exposure and vulnerability according to the quick assessment process, as defined in Chap. 4, Sect. 4.3. x-axes are expressed in hours of the day, and time-dependent values concern the users' outdoor density from an overall (all the present users) and an effectively exposed (ideally placed outdoors) standpoint (panel A), the users' normalized number and the ration between the exposed users and the whole number of users (panel B), the percentage of users by position and use behaviours (panel C), and age (panel D)

provide additional insights into the contribution of visitors to special buildings to improve the whole reliability of scenarios.

5.5 Simulation Scenarios and Results

Outcomes of Sects. 5.2, 5.3, and 5.4 have been then merged to provide emergency and evacuation scenarios to be assessed according to behavioural design simulations. Table 5.5 provides the full summary of these scenarios, which both include pre- (Sect. 5.2) and post-retrofit conditions (Sect. 5.3), and which are also characterized by different distribution of users at the start of the evacuation process, and two attack typologies (no attack conditions, thus implying simple OA evacuation, to have a baseline and reference scenario for comparisons, as discussed in Chap. 4, Sect. 4. 4; T2, according to relevance assessment discussed in Sects. 5.2 and 5.3). Table 5.5 also reports simulation codes then used in the following discussion of results.

Simulations have been performed according to the model defined in Chap. 3, Sect. 3.4, and mainly using validated tools developed under the BE S^{2}ECURe project [6]. In particular, the model has been implemented in Netlogo 6.2.0 [7]. Simulation results have been then analysed according to the behavioural KPIs reported in Chap. 4, Sect. 4.4.

Table 5.5 Shortlist of simulated scenarios in terms of users' distribution in the OA, attack typology and mitigation strategies, associated with the related simulation code used in the following discussion of results

Simulation code	Main rules for users' distribution in the OA	Attack typology	Mitigation strategies
H-No-Pre	Homogeneous in all over the OA	No attack	No, pre-retrofit scenario
F-No-Pre	Visitors are placed in front of them, thus being focused within the related SoRs (see Fig. 5.7)	No attack	No, pre-retrofit scenario
AS1-No-Pre	Visitors are mainly placed in the northern part of the OA, near the Loggia and the Palombaro, see Fig. 5.8	No attack	No, pre-retrofit scenario
AS2-No-Pre	Visitors are mainly placed in the Southern part of the OA, near the bars/restaurants and their dehors	No attack	No, pre-retrofit scenario
AS1-T2-Pre	Visitors are mainly placed in the northern part of the OA, near the Loggia and the Palombaro, see Fig. 5.8	T2	No, pre-retrofit scenario
AS2-T2-Pre	Visitors are mainly placed in the Southern part of the OA, near the bars/restaurants and their dehors, see Fig. 5.8	T2	No, pre-retrofit scenario
AS1-T2-ST1	Visitors are mainly placed in the northern part of the OA, near the Loggia and the Palombaro, see Fig. 5.9	T2	See Fig. 5.9
AS2-T2-ST1	Visitors are mainly placed in the Southern part of the OA, near the bars/restaurants and their dehors, see Fig. 5.10	T2	See Fig. 5.10
AS2-T2-ST2	Visitors are mainly placed in the Southern part of the OA, near the bars/restaurants and their dehors, see Fig. 5.10	T2	See Fig. 5.10

Figure 5.12 shows the evacuation curves for the simulated scenarios. In detail, Fig. 5.12 groups simulation curves by users' initial position and thus attack points without effects of the attack (simple evacuation) in pre-retrofit conditions, while Fig. 5.12b, c compare pre- (see Fig. 5.8) and post-retrofit (Figs. 5.9 and 5.10) scenarios, with and without attack effects, depending on the users' initial position and thus attack points.

The same rationale in comparisons is provided by Table 5.5, which shows the main KPIs provided in Chap. 4, Table 4.6, that are: the normalized evacuation time at the 95th percentile of arrived users—$TN95$ [-]; the normalized flows at the 95th percentile of arrived users—$FN95$ [-]; the normalized number of physical contacts among the users—PN [-]; the casualty ratio—CR [-]; and the not-arrived users' ratio—NA [-]. Selected comparisons are provided to evaluate the impact of the users' position on the overall evacuation process, and then to assess how the combination of attack typology and mitigation strategies can affect users' safety.

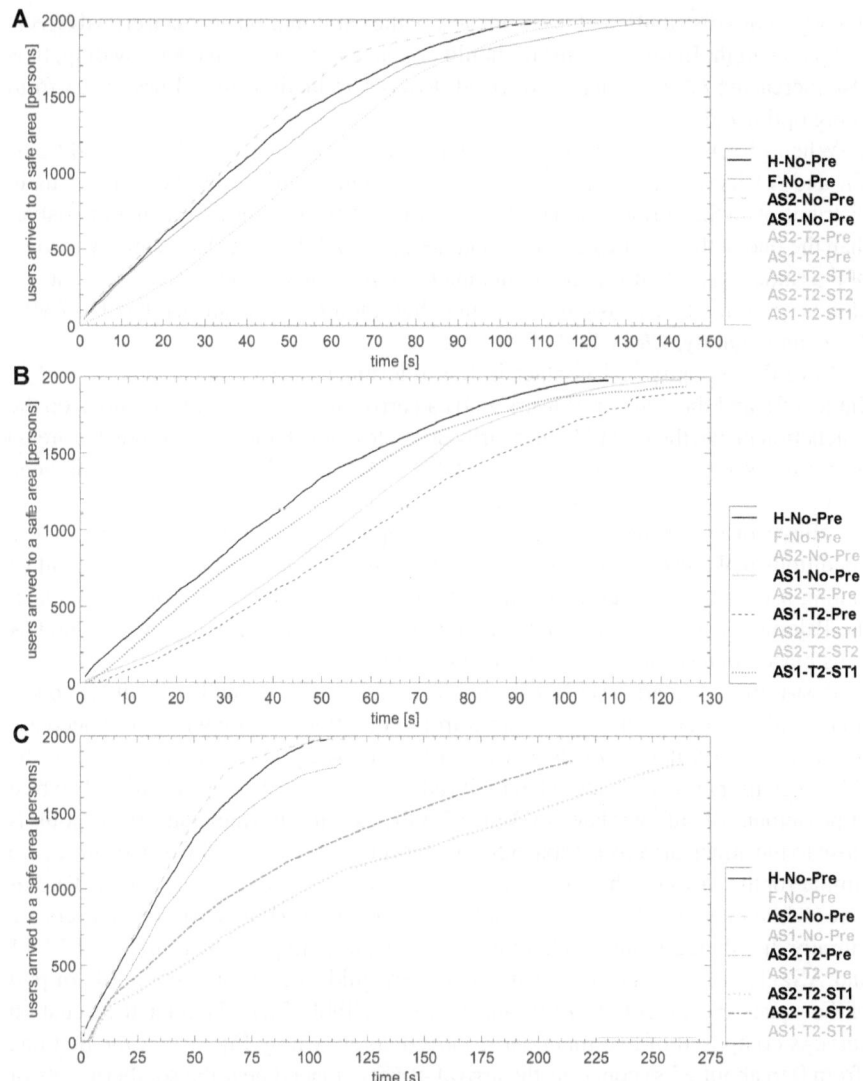

Fig. 5.12 Evacuation curves comparison in pre-retrofit scenarios, without attack effects (A), as well as in pre- and post-retrofit scenarios, with and without attack effects, considering the terrorist act involving the users placed in AS1 (B) and AS2 (C) parts of the square (see Figs. 5.8, 5.9 and 5.10). Simulation codes are reported in Table 5.5

In particular, in the scenarios without the attack effects on the crowd, and considering pre-retrofit conditions, the evacuation process seems to be quicker when users are placed in the southern part of the OA (AS1-No-Pre), since they are placed closer to square exits (Fig. 5.12a). As expected, when visitors are focused in front of the special buildings (F-No-Pre), *TN95* increases up to + 25% with respect to the scenario with a homogeneous distribution of users, in view of the combination between crowd effects and path length. In this case, users should organize movement in overcrowding, thus also increasing $FN95$, which corresponds to a risk reduction since flows are far from being optimized.

When effects of the attack are present, and considering pre-retrofit conditions, CR and NA increase, as expected. The scenario characterized by the visitors' main distribution in the northern part of the OA (AS1-T2-Pre) seems to be generally riskier than the one with users placed in the southern part (AS2-T2-Pre), essentially in view of the same issues on users' paths and interactions which can be noticed in no attack effects conditions (see positive Percentage Variation PV [%] values related to $TN95$, $FN95$ and, mainly, CR in Table 5.5).

Nevertheless, physical contact is less relevant (see PV [%] related to CR in Table 5.5), and the number of users who can arrive to a safe area and do not stop the evacuation inside the square (e.g., nearby obstacles, or in temporarily protected areas) increases, too (see PV [%] reduction related to NA in Table 5.5). These phenomena could be linked to the wider area in which users are initially placed as well as to the effects of the obstacles in the square (compare Fig. 5.5). In AS1-T2-Pre, these conditions make users organize evacuation, while, in AS2-T2-Pre, physical contact among users is relevant at the starting of the process, and the obstacles placed near the Palombaro represent a protection area while users are moving towards the OA exits placed in the northern part of the OA itself.

Post-retrofit scenarios achieve a significant decrease in casualties in all the conditions, with respect to the related pre-retrofit conditions, as shown by CR decrease in Table 5.5. Similarly, NA decreases when a LEA's point is implemented in the OA, since users are more attracted by it rather than by obstacle protection. The best improvement of safety relates to AS1-T2-ST1, essentially since the LEAs' point is close to the attack area (see Table 5.5). As expected, evacuation times and curves are similar for the cases in which users are initially placed near the Loggia and the Palombaro, while they significantly vary in the AS2 scenarios (Fig. 5.10) where users are initially distributed in the southern part of the square. In particular, in AS2-T2-ST1 and AS2-T2-ST2, supporting main emergency guidance towards the northern part of the square increases both $TN95$ and $FN95$ (see Table 5.5). The related evacuation curve is composed of two main parts, indeed, as shown by Fig. 5.12c: the first one (from 0 to about 25 s) concerns the arrival of users placed near the southern exits of the OA, while the second one (from 25 s) concerns the arrival of users towards the northern OA exits. Nevertheless, introducing the LEA's point in the square reduces the evacuation timing, as shown in Table 5.5 and Fig. 5.12c. Finally, it is worth noting that physical contact among users strictly depends on the initial position of the users when implementing mitigation strategies. In AS1-T2-ST1, PN increases, essentially in view of the great attractiveness of a unique point for the users, in the centre of the northern part of the OA. In both AS2-T2-ST1 and AS2-T2-ST2, on the contrary, PN decreases, essentially in view of the organized movement of users in a single direction, which reduces possible counterflow effects (Table 5.6).

Table 5.6 Key performance indicators by simulation code for evacuation risk assessment in case of terrorist acts in the OA, based on simulation results, according to Chap. 4, Sect. 4.4, and related Percentage Variation *PV* [%]. Comparisons are performed considering different groups of simulations with respect to their reference scenario (*ref*)

Simulation code		KPIs (PC)				
Scen	Ref	TN95 [-]	FN95 [-]	PN [-]	CR [-]	NA [-]
Pre-retrofit, without attack effects						
H-No-Pre		0.15 (*)	0.3 (*)	0.11 (*)	0 (*)	0 (*)
F-No-Pre	H-No-Pre	0.19 (25%)	0.44 (47%)	0.04 (-63%)	0 (n.a.%)	0 (n.a.%)
AS1-No-Pre	H-No-Pre	0.17 (14%)	0.39 (30%)	0.12 (10%)	0 (n.a.%)	0 (n.a.%)
AS2-No-Pre	H-No-Pre	0.14 (-7%)	0.25 (-16%)	0.14 (28%)	0 (n.a.%)	0 (n.a.%)
Pre-retrofit, with attack effects						
AS1-No-Pre	AS2-No-Pre	0.17 (9%)	0.43 (14%)	0.12 (-20%)	0.24 (24%)	0.04 (-49%)
Pre versus post-retrofit, without versus with attack effects						
…Comparing AS1						
AS1-No-Pre		0.17 (*)	0.39 (*)	0.12 (*)	0 (*)	0 (*)
AS1-T2-Pre	AS1-No-Pre	0.17 (-1%)	0.43 (11%)	0.12 (0%)	0.24 (n.a.%)	0.04 (n.a.%)
AS1-T2-ST1	AS1-T2-Pre	0.16 (-5%)	0.36 (-16%)	0.17 (42%)	0.08 (-68%)	0.02 (-50%)
…Comparing AS2						
AS2-No-Pre		0.14 (*)	0.25 (*)	0.14 (*)	0 (*)	0 (*)
AS2-T2-Pre	AS2-No-Pre	0.16 (11%)	0.38 (52%)	0.15 (8%)	0.19 (n.a.%)	0.08 (n.a.%)
AS2-T2-ST1	AS2-T2-Pre	0.4 (159%)	0.76 (100%)	0.06 (-60%)	0.17 (-12%)	0.08 (0%)
AS2-T2-ST2	AS2-T2-Pre	0.32 (107%)	0.7 (85%)	0.06 (-60%)	0.16 (-19%)	%1.%2 -12%

References

1. Rota L (1990) Matera: storia di una città. Edizioni Giannatelli
2. D'Imperio N, Giase F (2017) Conoscere matera. Edizioni Magister, Matera
3. Cantatore E, Quagliarini E, Fatiguso F (2024) Terrorism Risk Assessment for Historic Urban Open Areas. Heritage 7:5319–5355. https://doi.org/10.3390/heritage7100251
4. The European Commission (2022) Security by design: protection of public spaces from terrorist attacks
5. Beňová P, Hošková-Mayerová Š, Navrátil J (2019) Terrorist attacks on selected soft targets. J Secur Sustain Issues 8:453–471. https://doi.org/10.9770/jssi.2019.8.3(13)

6. Quagliarini E, Bernardini G, D'Orazio M (2023) How could increasing temperature scenarios alter the risk of terrorist acts in different historical squares? a simulation-based approach in typological Italian squares. Heritage 6:5151–5188

7. Wilensky U, Rand W (2015) An introduction to agent-based modeling. Modeling Natural, Social, and Engineered Complex Systems with NetLogo. MIT Press

Chapter 6
Conclusions and Perspectives

Abstract Risk assessment and mitigation again terrorist acts in outdoor Open Areas (OAs) should be based not only on the analysis of possible hazard, physical vulnerability, and perpetrator behaviours and "modus operandi", but also on joint investigation of the user exposure, vulnerability, and behaviours in emergency conditions. A behavioural design approach relying on the analysis of emergency and evacuation via simulation tools could support these tasks, since it is able to represent complex interactions among these factors and to include users' reaction and needs to the terrorist event. Indeed, this approach should be supported by quick methods for scenario creation, balancing efforts to manage reliable data and to determine critical phenomena in the OAs. This book offers the definition of a risk assessment and mitigation methodology according to such an approach, applying it to a relevant real-world OA to demonstrate its capabilities in supporting local authorities and their technicians in facing terrorist acts in (over)crowded situations. Different conditions in attack points, users' exposure and vulnerability and implemented mitigation strategies are derived and tested through simulations, allowing to determine users' risk levels in emergency and evacuation depending on the combination of such inputs. Nevertheless, the capabilities of the methodology and of its tools should be extended, thus needing further efforts related to both research and practice. This chapter first traces an overview of the main objectives and then suggests future directions for this approach development and application according to both researchers and decision-makers' standpoints.

Keywords Behavioural design · Terrorist act · Risk assessment · Risk mitigation · Outdoor open areas

G. Bernardini et al., *Terrorist Risk in Urban Outdoor Built Environment*,
SpringerBriefs in Architectural Design and Technology,
https://doi.org/10.1007/978-981-97-6965-0_6

6.1   Outdoor Open Areas and Terrorist Acts: How Behavioural Design Could Support Risk Assessment and Mitigation?

The significance of outdoor Open Areas (OAs) in the context of terrorist threats is essentially due to their characterizing features. First, a high level of desirability by perpetrators, which can be also expressed in likelihood terms, is widely correlated to the possible (over)crowding conditions that can be hosted by OAs, in view of their paramount rule and attraction in the whole urban Built Environment (BE). In this sense, OAs can also widely host buildings with special intended uses and emblematic functions, such as worship places, cultural areas and government buildings, increasing the "visibility" of attack consequences and the related symbolic value [1]. Moreover, they are "soft targets", which generally implement a low level of structured measures for protection and mitigation of possible attacks [2], since they essentially are "ordinary public places" in the urban BE and "hard" strategies could limit the possibility of fruition by communities. In that sense, they can be also characterized by critical situations in terms of users' vulnerability, depending on the users' age, gender, familiarity with the BE, awareness of proper response in case of an attack, motion and sensory abilities [3]. Consequences of attack could then have a wide impact on the community, provoking medium to extreme effects in terms of casualties, as confirmed by statistics related to the European context [4].

In view of these risk factors, risk assessment tasks are essential to make local authorities aware of possible scenarios to be faced and to then select and deploy Risk Mitigation and Reduction Strategies (RMRSs) [5]. Codified regulations have been proposed in the last few decades by risk-prone countries, and wide applications in real-world scenarios have been provided, but many problems seem to be still present in both risk assessment and mitigation. In particular, risk assessment and the definition of RMRSs seem to be widely based on standardized issues, which do not consider, for instance, that users' exposure and vulnerability can vary over time. Perpetrators' behaviours should be better related to the OAs features, especially while detecting possible points of attack where to focus solutions. Users' response in emergency conditions is generally assessed in a deterministic manner, assuming the same behaviours for all the exposed individuals, and "standard" reactions which are often derived from other kinds of emergencies, especially in relation to the immediate evacuation process (i.e., from fire emergency behaviours). Moreover, RMRSs should be also defined depending on the OAs features and the exposed users, to properly mix structural (e.g., protective barriers, building components, space design, control systems) and non-structural (e.g., emergency management and planning, users' preparedness and awareness) solutions.

A holistic standpoint should be introduced to fully evaluate the combination of perpetrator behaviours and user behaviours depending on the OAs morphology, vulnerability and constructive features. The behavioural design approach could move in this direction [6]. This kind of approach considers that risk assessment should be based on effective users' features and behaviours, thus also depending on the way the OA is used by the crowd over space and time, and on the possible interactions among them, the perpetrators and the physical scenario. Therefore, the effectiveness of RMRSs should be assessed according to the same behavioural perspective, to determine which scenarios could be more relevant for users and how different strategies can support users in emergency and evacuation. Methods for scenario creation can support these goals, combining OA features, users' exposure and vulnerability, and perpetrators' will and "modus operandi". Moreover, simulation-based techniques are also encouraged to analyse the interactions of such factors. In that sense, the behavioural design also extends methods already adopted in other contexts, such as those of fire safety in buildings [7], e.g., those based on "Psychonomics" [8], but adopts specific risk-affecting features and behaviours strictly related to terrorist acts rather than adopting related general purposes or "out-of-context" ones.

This book provides an overview of the behavioural design approach in the context of terrorist acts in OAs, by tracing the different issues with respect to the event phenomenology, the OA features, the user-oriented factors and RMRSs, and taking advantage of both quick and analytical tools. The whole methodology allows research to define a clear overview of the advances on the matter, but also supports local authorities and decision-makers in the process of knowledge, planning and mitigation of terrorist risks in their OAs. In particular, the proposed behavioural design approach succeeds in:

1. Determining a risk matrix correlating consequences and likelihood levels according to a frequentistic standpoint, thus relying on the analysis of terrorist threats in correlation to real-world conditions and previous events (Chap. 2). This matrix can mainly support decision-makers in identifying outcoming risk levels depending on the typology of the OA and the attack.
2. Classifying codified RMRSs according to integrated criteria for their design and application, that are redundancy, coordination, adaptability, application context, correlation with (over)crowding and costs (Chap. 2). Such results can support researchers in having a structured overview of solutions from a multi-perspective level, as well as local authorities in evaluating possible constraints for their applications in specific case studies.
3. Defining effective users' behaviours in emergency and evacuation due to terrorist acts, by providing a simulation model based on an agent-based approach and describing also related motion quantities (Chap. 3). Researchers can use the structured list of behaviours to check their relevance in real-world events, thus adding or modifying the statistical frequency of the detected ones. Similarly, density-speed correlations can be updated in the future by researchers, as well as used in evacuation simulation modelling. Moreover, agent-based models can be defined according to the general rules provided in this work, evaluating the impact of alternative behavioural patterns on the evacuation process too. These models can be then used by decision-makers to explore users' response in real-world scenarios.
4. Providing methods for scenario creation, which consider both hazard, OA vulnerability, users' exposure, and vulnerability (Chap. 4). In particular, methods allow

to determine: (1) possible points of attack depending on the desirability, the physical features characterizing the OA and its layout, the possible countermeasures, and the intended uses of the OA, which also relates to both occupancy and perpetrators' desirability; (2) time-dependent assessment of users' exposure (number of exposed users) and vulnerability (typology by age, gender) within the OA. In this sense, these methods pursue a quick application approach, using remote surveys and online databases, but they could be also supported by in situ surveys to increase the reliability of results. Decision-makers can use these methods to provide bases for risk assessment, as also demonstrated thanks to the case study application (Chap. 5);

5. Defining key performance indicators (KPIs) for risk assessment based on users' response to emergency evacuation in case of a terrorist act (Chap. 4). KPIs summarize the impact of user-perpetrators-OA interactions taking into account the different effects of the event on the crowd, thanks to the application of evacuation simulators. Normalized indicators are provided, so as to make them comparable since they range in the same interval (0 to 1, as maximum impact on the users). Decision-makers can use these KPIs to compare several scenarios comprising different typologies of attack in the same OA, different levels of users' exposure and vulnerability, as well as different implemented RMRSs. Nevertheless, KPIs are not dependent on the evacuation simulation criteria; thus, different tools or models can be applied by then investigating results using the same KPIs. The application to the case study (Chap. 5) demonstrates how these KPIs can point out different phenomena related to the users' evacuation and to determine if a certain RMRS can support some of them.

Moreover, the matrices, simulation-based KPIs, and assessment methods about the point of attack and the user-related factors can be used not only with reference to different conditions in the same OA, including pre- and post-retrofit scenarios. In fact, they can compare and contrast risk conditions in diverse OAs within the same urban BE. Considering a pre-retrofit context, they can hence make local authorities aware of the OAs where terrorist acts can provide higher impact and that should be the object of specific interventions. Considering pre- versus post-retrofit contexts, they can also suggest which OAs can take the larger benefits from RMRSs implementation.

6.2 Perspectives in Research and Practice

The contribution to terrorist act risk assessment and mitigation given by the behavioural design approach relies on the user-centred and experimental-based criteria on which it is defined, and the case study application demonstrates the capabilities of the tools provided in this book. Nevertheless, specific perspectives can be associated with the goals reached.

6.2.1 Risk Matrix

The definition of risk matrix provided in Chap. 2 results from the phenomenological analysis of terroristic events that occurred in Western Europe in the urban BE and recorded in the GTD database. The use of a properly-structured database allowed to understand the phenomenon according to a well-thought parametrization of the BE itself, aiming at the quantification of the relevance of OAs as soft target. This has been achieved not only by assessing events that occurred in streets and squares, but also by analysing all the events that occurred outside the buildings. The assessment of recurrences and consequences of such events has demonstrated how the proneness of OAs is strictly related to the use of outdoor open areas as systems of buildings, uses and infrastructures, also considering the most efficacious and recurrent attack types.

In that sense, risk matrices applied to all the BE represent a rapid tool to support decision-makers in determining the potential riskiness of some places and the main weapons and modus operandi to focus on, also taking into account or borrowing previous detailed experiences (regulation, guidelines). On the other hand, these levels of detail can enhance global awareness of risk and emergency scenarios, preparing detailed solutions (such as mitigative and preventive strategies) properly focused on the elements and features that are part of such BE parts and may alter the risk conditions [9, 10]. From such considerations, the parameterization of the OAs has arisen in terms of characters and properties of physical space and items, trying to simplify the real specifications of OAs towards the identification of factors that affect the hazard, vulnerability and/or exposure, as the basis of a critical reading of RMRSs and their efficacy to the prevalent attack types, and the setting up of an expeditious way to assess the risk of real OAs.

Overcoming the advantages and results already reached in this study, some considerations can be highlighted in order to improve the overall issue.

The first observation relates to the static nature of risk matrices; terroristic events are strongly related to the decision of perpetrators which can interfere with the final riskiness of places also considering the variable conditions within an OAs. In fact, the risk proneness in some classes of BE may change in consideration of their prolonged or limited time of use (e.g., some hours in a day). In that sense, the risk matrices require to be implemented with time-related information which can enhance the global riskiness, towards three-dimensional and time-related risk matrices [11, 12].

The second point of discussion can be related to the consequence level. The matrix used in the methodology derives from the assessment of risk considering the human exposure (number of fatalities and victims) while the breakdown of services and/or physical damages are neglected [13, 14]. A possible economic-related assessment can enhance the final tool, taking into account previous major events to discuss the relevance of the economic dimension in the assessment, without neglecting the attack type and the modus operandi.

6.2.2 Behavioural Modelling and Simulation

Detected behaviours in the case of terrorist acts defined in Chap. 3 are based on the analysis of literature works mainly focused on the European context. The statistical validity of data could be hence affected by limitations in view of the specific geographical area where the event could happen, as well as by the sample dimensions in the referenced works. Nevertheless, they are essentially connected with the limited extent of research on such themes, and thus on the limited analysis of real-world attacks, although the availability of videotapes that can be used to examine the users' response in an almost unbiased manner. In this sense, researchers should focus their effort on the extension of current sample database dimensions, considering the widest number of scenarios in terms of attack typology and geographical area.

Moreover, issues related to individual vulnerability should be better codified, thus examining how age, gender, and motion abilities can impact the selection of specific behaviours. These data could be also supported by traditional investigation methods, such as those based on surveys (both hypothetical scenarios and involving survivors of real-world accidents). Moreover, the analysis of real-world events could be coupled with virtual reality studies [15], although their complete reliability should be fully demonstrated. The same virtual reality works could be also used to support training actions for preparedness and awareness increase of users against terrorist acts, thus connecting researchers and stakeholders' aims.

Similarly, the analysis of first responders' behaviours, including interventions of law enforcement agencies (LEAs), should be improved, along with those on perpetrators' actions, so as to fully define the simulation models including useful details which can be represented using the proposed agent-based approach. These data can be used to perform additional verification and validation of models, following consolidated approaches used in the fire safety field [16], and also using real-world scenarios reconstructed by simulators.

From a modelling perspective, finally, the present work model and application rely on the combination of agent-based modelling with cellular automata [17]. Other modelling approaches, including microscopic ones (e.g., force-based), could be used to implement the same evacuation rules, also enriching them with additional behaviours of other agents (i.e., LEAs and perpetrators) which can also depend on the "modus operandi" as well as on additional environmental conditions. For instance, approaches based on a continuous representation of the movement space could improve the level of detail of the results, as well as trace microscopic phenomena to be assessed by KPIs.

6.2.3 Scenario Creation and KPI-Based Risk Assessment

Methods for the creation of input scenarios for simulations provided in Chap. 4 rely on quick approaches, essentially based on the definition of main descriptors of

OA, users and attack, and exploiting remote access sources. Codified frameworks to implement in situ surveys could be hence developed, by also providing standard forms which can be used by decision-makers to assess these elements within their application scenarios.

Method for the identification of the point of attack takes advantage of expeditious risk calculations determined for the qualification of real OAs case study exposed to terroristic attacks and properly set for the armed assault (T2) and bombing with vehicles (T3) [18]. Basis of the data collection is represented by the morphology of OAs, the presence of the main extension of obstacles within the outdoor area, and the intended uses and dimensions of buildings. All these data can be gathered using available details. However, symbolicity of places and buildings, as well as their economic, cultural, and political significance at the local, national and/or international scales requires to be studied when not universally known. The method is also supported by final risk matrices declined to OAs proneness to T2 and T3 attack types, as a fast descriptive tool for the qualification of partial and/or global risk level (high, medium, low, negligible) of OAs. In fact, the risk calculation algorithm can be declined into a risk property of the OAs and their parts, thanks to the use (in geometry and position) of the Space of Relevance. This is intended as the external areas located along the façades of buildings with the dominant use for the OAs riskiness (public or commercial) or significance (symbolic or strategic), calculated coherently with the internal function and maximum intended density.

Methods for users' exposure and vulnerability assessment from a time-dependent perspective are based on the identification of space types in the OA, time occupation and standard users' density, and enable the analysis of trends in OA use dynamics. The same method can be applied for working days and holidays, but also for specific events (e.g., mass gatherings) and for different seasonal periods (e.g., summer versus winter). The current methodology relies on a conservative approach since the maximum capacity of occupation is assumed according to current regulations (i.e., those related to fire safety). Nevertheless, the methods do not take into account specific additional conditions which can vary the user-related factors, such as those related to environmental quality and climate/meteorological data. Nevertheless, these factors can vary the presence of users in a given space [19]. For instance, outdoor temperatures can alter the areas where users gather and perform leisure activities, especially in the hottest period of the year. Similarly, additional temporary uses within the OA should be added by the methodology, moving towards a framework to integrate in situ analysis of effective users' behaviours (e.g., rapid and remote analysis can suggest users are focused on a certain part of the OA, but effective conditions are different from theoretical; pedestrian flows in crowded areas). In that sense, local administrations should balance the quick assessment capabilities with their specific level of knowledge on the effective OA use by citizens. Nevertheless, indicators and the overall workflow proposed by this book could be not changed by these detailed analyses.

KPIs provided by this book cover a variety of effects of terrorist acts on the users, and thus can be reliably used to describe risk levels in both pre- and post-retrofit scenarios. Nevertheless, additional KPIs can be included to describe further specific issues of interest, especially if relating to microscopic dynamics. In fact, the proposed KPIs describe the overall effects of the attack on the scenario, or even on the crowd as a whole, depending on microscopic interactions, but local phenomena are not assessed by them. In this sense, KPIs could be moved into risk maps, e.g., on the spatial distribution of users' casualties and physical contact, on density, on users' paths with respect to protective obstacles. Moreover, KPIs can be combined into risk metrics to synthetically determine terrorist act risk according to a unique indicator, which could be used to rank different scenarios of the same OA and/or different OAs. A unique metric-based indicator could support decision-makers in risk assessment by limiting complexities due to the interpretation of specific composing KPIs. In this case, KPIs can be also balanced depending on their effective weight on the final risk for users. Therefore, additional research efforts should be still provided towards such a direction.

Finally, this book provides an application to a real-world case study represented by a significant OA. Further applications to other OAs are needed to fully demonstrate the capabilities of the behavioural design approach, but verification tasks should be also carried out to support the reliability of the model by comparing outputs of the method application and conditions in real-world emergencies. It is worth noting that the model and framework provided in this book are based on phenomenological and experimental-based data, according to the general criteria of the behavioural design approach. Moreover, the behavioural evacuation simulation tool used in Chap. 5 for the case study application [17] has been verified according to consolidated testing rules [20]. Nevertheless, applying the proposed KPIs to data related to real-world terrorist acts would be useful to additionally compare and contrast possible differences in scenario definition (Chap. 4, Sects. 4.2 and 4.3) as well as in simulation outcomes (Chap. 3, Sect. 3.4). Similar efforts have been provided by other works [21], although they seem to be limited in terms of scenarios and, mainly, to have poor relevance with OA events. A possible lack of structured data on these scenarios from a complete perspective should be then solved by future works, too.

6.2.4 Risk Mitigation

Risk mitigation should be supported by the analysis of users' behaviours in emergency conditions, as discussed above, but also by a proper level of knowledge of possible strategies by decision-makers. The paradigms related to coordination, redundancy, adaptability, and costs provided in this work can then take advantage of the analysis of OA users' support in evacuation, pursuing a holistic perspective and moving towards sustainability against terrorist acts. In this sense, further efforts are needed to provide behavioural-based and specific analysis on the effectiveness of mitigation strategies, thanks to simulations and/or the assessment of their impacts in real-world events, or virtual reality scenarios. This kind of action could move towards the use of simplified tables to guide the selection of mitigation strategies, that could then be tailored and tested again using evacuation simulators. Typological

OAs can be used to this end, considering that they are archetypes of real-world OAs and that they rely on their main common features.

At the same time, this book provides an application to risk mitigation in a specific case study by considering emergency management strategies according to the identity features of the historical application scenario. Moreover, these RMRSs can be easily implemented in each environment, being supported, for instance, by wayfinding signs, deployment of first responders and, partially, users' preparedness and awareness campaigns. In this sense, they have also a limited cost in terms of physical implementation and could be easily modified to face different events. Therefore, further efforts should also move towards the combination of this kind of RMRSs with structural solutions, which will be developed, designed, and implemented depending on the specificities of the considered attack typologies and on the OA layout too.

References

1. Kalvach Z, et al. (2016) Basics of soft targets protection—guidelines (2nd version). Prague
2. The European Commission (2022) Security by design: protection of public spaces from terrorist attacks
3. Quagliarini E, Bernardini G, Romano G, D'Orazio M (2023) Users' vulnerability and exposure in public open spaces (squares): a novel way for accounting them in multi-risk scenarios. Cities 133:104160. https://doi.org/10.1016/j.cities.2022.104160
4. Cantatore E, Quagliarini E, Fatiguso F (2022) European cities prone to terrorist threats: phenomenological analysis of historical events towards risk matrices and an early parameterization of urban built environment outdoor areas. Sustainability 14:12301. https://doi.org/10.3390/su141912301
5. Quagliarini E, Fatiguso F, Lucesoli M et al (2021) Risk reduction strategies against terrorist acts in urban built environments: towards sustainable and human-centred challenges. Sustainability 13:901. https://doi.org/10.3390/su13020901
6. Bernardini G, Quagliarini E (2021) Terrorist acts and pedestrians' behaviours: first insights on European contexts for evacuation modelling. Saf Sci 143:105405. https://doi.org/10.1016/j.ssci.2021.105405
7. Ronchi E, Corbetta A, Galea ER et al (2019) New approaches to evacuation modelling for fire safety engineering applications. Fire Saf J 106:197–209. https://doi.org/10.1016/j.firesaf.2019.05.002
8. Kobes M, Helsloot I, de Vries B, Post JG (2010) Building safety and human behaviour in fire: a literature review. Fire Saf J 45:1–11. https://doi.org/10.1016/j.firesaf.2009.08.005
9. Al Amosh H, Khatib SFA, Ananzeh H (2024) Terrorist attacks and environmental social and governance performance: evidence from cross-country panel data. Corp Soc Responsib Environ Manag 31:210–223. https://doi.org/10.1002/csr.2563
10. Amin F, Verma K, Acharya P (2024) Multi-hazard risk and integrated approach to resilience. In: Disaster risk and management under climate change. Springer, pp 581 592
11. Amirshenava S, Osanloo M (2018) Mine closure risk management: an integration of 3D risk model and MCDM techniques. J Clean Prod 184:389–401
12. Bao C, Li J, Wu D (2022) Three-dimensional risk matrix: theoretical basis and construction. In: Risk matrix: rating scheme design and risk aggregation. Springer, pp 149–169
13. Tan W, Wang W, Zhang W (2024) The effects of terrorist attacks on supplier–customer relationships. Production and Operations Management 10591478231224920

14. Šarūnienė I, Martišauskas L, Krikštolaitis R et al (2024) Risk assessment of critical infras-
    tructures: a methodology based on criticality of infrastructure elements. Reliab Eng Syst Saf
    243:109797
15. Lovreglio R, Ngassa D-C, Rahouti A, et al (2021) Prototyping and testing a virtual reality
    counterterrorism serious game for active shooting. SSRN Electron J. https://doi.org/10.2139/
    ssrn.3995851
16. Ronchi E (2021) Developing and validating evacuation models for fire safety engineering. Fire
    Saf J 120:103020. https://doi.org/10.1016/j.firesaf.2020.103020
17. Quagliarini E, Bernardini G, D'Orazio M (2023) How could increasing temperature scenarios
    alter the risk of terrorist acts in different historical squares? a simulation-based approach in
    typological Italian squares. Heritage 6:5151–5188. https://doi.org/10.3390/heritage6070274
18. Cantatore E, Quagliarini E, Fatiguso F (2024) Terrorism Risk Assessment for Historic Urban
    Open Areas. Heritage 7:5319–5355. https://doi.org/10.3390/heritage7100251
19. Choi Y, Yoon H, Kim D (2019) Where do people spend their leisure time on dusty days?
    application of spatiotemporal behavioral responses to particulate matter pollution. Ann Reg
    Sci 63:317–339. https://doi.org/10.1007/s00168-019-00926-x
20. Ronchi E, Kuligowski ED, Reneke PA, et al (2013) The process of verification and validation
    of building fire evacuation models. NIST Technical Note 1822:
21. Li S, Zhuang J, Shen S (2017) A three-stage evacuation decision-making and behavior model
    for the onset of an attack. Transp Res Part C: Emerg Technol 79:119–135. https://doi.org/10.
    1016/J.TRC.2017.03.008